D0477445

WETLAND SYSTEMS TO CONTROL URBAN RUNOFF

WETLAND SYSTEMS TO CONTROL URBAN RUNOFF

MIKLAS SCHOLZ

The University of Edinburgh
Edinburgh, UK

ELSEVIER

Amsterdam • Boston • Heidelberg • London • New York • Oxford
Paris • San Diego • San Francisco • Singapore • Sydney • Tokyo

Elsevier
Radarweg 29, PO Box 211, 1000 AE Amsterdam, The Netherlands
The Boulevard, Langford Lane, Kidlington, Oxford OX5 1GB, UK

First edition 2006

Copyright © 2006 Elsevier B.V. All rights reserved

No part of this publication may be reproduced, stored in a retrieval system
or transmitted in any form or by any means electronic, mechanical, photocopying,
recording or otherwise without the prior written permission of the publisher

Permissions may be sought directly from Elsevier's Science & Technology Rights
Department in Oxford, UK: phone (+44) (0) 1865 843830; fax (+44) (0) 1865 853333;
email: permissions@elsevier.com. Alternatively you can submit your request online by
visiting the Elsevier web site at http://elsevier.com/locate/permissions, and selecting
Obtaining permission to use Elsevier material

Notice
No responsibility is assumed by the publisher for any injury and/or damage to persons
or property as a matter of products liability, negligence or otherwise, or from any use
or operation of any methods, products, instructions or ideas contained in the material
herein. Because of rapid advances in the medical sciences, in particular, independent
verification of diagnoses and drug dosages should be made

Library of Congress Cataloging-in-Publication Data
A catalog record for this book is available from the Library of Congress

British Library Cataloguing in Publication Data
A catalogue record for this book is available from the British Library

ISBN-13: 978-0-444-52734-9
ISBN-10: 0-444-52734-6

For information on all Elsevier publications
visit our website at books.elsevier.com

Printed and bound in The Netherlands

06 07 08 09 10   10 9 8 7 6 5 4 3 2 1

Working together to grow
libraries in developing countries

www.elsevier.com | www.bookaid.org | www.sabre.org

ELSEVIER   BOOK AID International   Sabre Foundation

Acknowledgements

I would like to thank all the current and previous members of my Urban Water Research Groups at The University of Edinburgh and the University of Bradford for their research input. Particular thanks go to Mr. Byoung-Hwa Lee, Mrs. Sara Kazemi-Yazdi and Mrs. Jing Xu for their outstanding contributions to past research papers.

I would like to acknowledge the institutions that provided funding for my research; in particular, The Royal Academy of Engineering, Glasgow City Council and the Alexander von Humboldt Foundation.

I am also grateful for the support received from the publishing team at Elsevier.

I would like to dedicate this book to my family who supported me during my studies and career.

Particular thanks go to my wife **Natascha**, my children **Philippa** and **Jolena**, and my mother **Gudrun**.

About the Author

Miklas Scholz, cand ing, BEng, PgC, MSc, PhD, ILTM, CEnv, CSci, CEng, MCIWEM, MICE, FIEMA, is the author of this particularly timely book entitled 'Wetland Systems to Control Urban Runoff'. He is a Lecturer (Grade B) in Civil and Environmental Engineering in the Grade 5-rated Institute for Infrastructure and Environment at The University of Edinburgh (Scotland). Previously, he held a Lecturership in Environmental and Water Engineering in the Department of Civil and Environmental Engineering at the University of Bradford (England).

He was recently awarded a Global Research Award (assessment of hydraulics of drainage ditches in Germany) and an Industrial Secondmentship (sustainable urban drainage system (SUDS) management in Glasgow (Scotland), hosted by Glasgow City Council) by The Royal Academy of Engineering. Dr Scholz also received research funding as a principal investigator from various other bodies including the Natural Environment Research Council, Alexander von Humboldt Foundation (Germany), Glasgow Council, Scottish Executive, Chinese Education Authority, The University of Edinburgh Development Trust, Atlantis Water Management, Formpave and Triton Electronics.

Dr Scholz's overall research interests are in urban water management with a particular passion for solving water quality problems associated with constructed treatment wetlands and ponds, SUDS technology and planning, and biological filtration and sludge management. Since 1999, he led more than ten substantial research projects comprising the following three key research areas:

Wetlands

- Constructed treatment wetlands: Dr Scholz assessed the treatment efficiencies of constructed wetlands, and proposed novel design, operation and water quality monitoring guidelines. The appreciation for flow processes through saturated and unsaturated porous media, biochemical treatment processes and microbial dynamics related to public health issues are relevant for this book.
- Natural wetlands: He came up with novel assessment and management guidelines with particular emphasise on the interaction between hydraulics, water quality and vegetation.

Sustainable water management

- SUDS: He developed advanced design, operation, management and water quality monitoring guidelines, particularly for attenuation and infiltration ponds, and basins. He also showed that the common goldfish can be used to increase public acceptance of SUDS, and to reduce algal growth. Transferable design and operational skills, as well as in-depth understanding of biochemical and physical processes impacting on water quality were gained.
- Treatment of surface water runoff and gully pot liquor: He proposed new design and operation guidelines for constructed wetlands treating urban water contaminated with heavy metals, and modelled systems with case-based reasoning and neural networks.
- Glasgow, Edinburgh and Penicuik SUDS management projects: Development of methodologies for SUDS implementation, design and management.

Filtration and sludge management

- Biological filtration systems: He improved treatment performances of various systems including sand and gravel filters, biological activated carbon filters and advanced filtration systems. Studies of fluid flow through porous media are relevant for this book.
- Membrane filtration technologies: He solved biofouling problems, and demonstrated that alternative low technology and affordable solutions can replace some novel (bio-)membrane technologies.
- Sludge dewaterability: Dr Scholz assessed dewaterability technologies, and proposed modifications to capillary suction time testing equipment that would allow numerical modelling for the first time.

Dr Scholz covers various subject disciplines in engineering and science including urban water engineering, management and modelling, civil and environmental engineering, biotechnology, chemical process technology and environmental science.

The author's national research reputation in urban water technology and management is outstanding, because he introduced advanced design and management guidelines to the environmental engineering community. He has also a very good international reputation in SUDS (particularly infiltration systems, ponds and especially wetlands), biochemical processes within filtration systems, and sludge dewaterability testing, because of his unconventional multi-disciplinary and holistic approach in solving engineering problems, which benefits the general public.

Public interest in his research is remarkable. For example, he has been interviewed by the journal *Nature*, and newspapers such as *The Observer*, the *Sunday Herald*, *The Sunday Post* and *Iran Daily*.

His reputation is predominantly based on 40 journal paper publications since 2001, and more than 40 presentations at national and international conferences. He has approximately 30 journal papers and 40 other publications that are directly relevant to this book. His strong publication and citation record is evidence for the interest of the wider research community in his research benefiting engineering and quality of life.

Preface

This novel, timely and cost-competitive book on 'Wetland Systems to Control Urban Runoff' covers water and environmental engineering aspects relevant for the drainage and treatment of stormwater, wastewater and contaminated natural watercourses from predominantly urban areas, providing a descriptive overview of the complex 'black box' treatment systems and design issues involved.

The fundamental science and engineering principles of relevant water and wastewater treatment units and processes are explained to address the student as well as the professional market. Standard and advanced design recommendations for predominantly constructed wetlands and related sustainable urban drainage systems (best management practice) are provided to account for the interests of the professional engineers and environmental scientists. Latest research findings in urban water management technology are discussed to attract both research- and teaching-active academics who should recommend this textbook to undergraduate and postgraduate students.

The book deals comprehensively not only with the design, operation, maintenance and water quality monitoring of traditional and advanced wetland systems, but also the analysis of asset performance and modelling of treatment processes of traditional and novel infrastructure predominantly in developed as well as developing countries, and the sustainability and economic issues involved.

The textbook is essential for undergraduate and postgraduate students, lecturers and researchers in the civil and environmental engineering fields of water engineering, public health engineering, and related non-engineering disciplines including environmental science. It is a useful reference for engineers and scientists working for the water industry, non-governmental organisations, local authorities and governmental bodies in the design and operation of wetland systems. Moreover, consulting engineers can apply practical design recommendations, and refer to a large variety of experimental research case studies.

The basic scientific principles are also of interest to all concerned with the urban environment, including town planners, developers, engineering technicians, public health workers and ecologists. The book has been written for a wide readership but sufficiently 'hot and sexy' research topics are also addressed to guarantee a wide public interest and a longer shelf-life.

Solutions to pressing water quality problems associated with natural and constructed treatment wetlands and ponds, and other sustainable biological filtration and treatment technologies linked to public health engineering are explained. Case study topics are diverse: wetlands including natural wetlands and constructed treatment wetlands; sustainable water management including sustainable drainage systems and best management practice. The research projects are multi-disciplinary, holistic and predominantly experimental. However, case-based reasoning and neural network modelling projects involving wetlands are also discussed.

Novel design and management guidelines for treatment wetlands are introduced to the engineering community. The content is biased towards biochemical processes within constructed treatment wetlands and similar systems such as ponds and infiltration basins. Moreover, novel and unconventional approaches in solving engineering problems are presented; e.g., integration of wetland systems into conventional water and wastewater treatment trains.

The book is predominantly based on experience gained by the author and his Urban Water Research Groups over the last ten years. Original material published in more than 30 high-ranking journal articles and 40 conference papers was revisited. Experience gained as an active editorial board member of several journals guarantees that this textbook contains a wealth of material that fills gaps in knowledge and understanding, and documents the latest cutting edge research in wetland systems to treat urban runoff.

This book provides material for readers with different levels of knowledge and understanding; e.g., background information for students, novel design aspects for professionals, and latest research findings for academics. The author hopes that this book will be an indispensable tool for the wetland community for many years to come!

The book tries to integrate natural and constructed wetlands, and sustainable drainage techniques into traditional water and wastewater systems used to treat surface runoff and associated diffuse pollution. Chapters 1–5 introduce water quality management, and water and wastewater treatment fundamentals to the inexperienced reader.

Chapters 6–11 review preliminary and predominantly primary treatment units that can be combined with wetland systems. Chapters 12–18 summarise predominantly secondary but also tertiary treatment technologies that can be used in combination with wetland technologies or as alternatives in cases where land availability is restricted due to costs. Usually non-essential traditional technologies are briefly presented in Chapters 19 and 20 for the reason of completeness.

Microbiological and disinfection issues relevant for treatment wetlands are covered in Chapters 21 and 22. Chapter 23 highlights sludge treatment and disposal options that should be considered for sludges from wetland systems.

Chapters 24–31 focus predominantly on timely research case studies related to constructed wetlands and associated technologies for runoff and diffuse pollution treatment. These chapters are written for professionals and students interested in design, process and management details.

Miklas Scholz
Edinburgh, 1 May 2006

PRAST	=	Prevalence Rating Approach for SUDS Techniques
QR	=	quantisation error
R	=	(*mean product moment*) correlation coefficient
R^2	=	coefficient of determination
RBC	=	rotating biological contactor
RBF	=	radial basis function
SD	=	standard deviation
SOM	=	self-organising map
SS	=	(*total*) suspended solids (mg/l)
SUDS	=	sustainable (*urban*) drainage system
SVM	=	support vector machine
T (*or t*)	=	infiltration time (s) or temperature (°C)
TE	=	topographic error (*usually in %*)
TOC	=	total organic carbon (mg/l)
TS	=	total solids (mg/l)
UK	=	United Kingdom
U-matrix	=	unified distance matrix
USA	=	United States of America
UV	=	ultraviolet (light)
WTW	=	Wissenschaftlich Technische Werkstätten (*German company*)
Z_1	=	factor (*defined by the BRE method*)
Z_2	=	growth factor (*defined by the BRE method*)

Contents

Chapter 1

Water quality standards

1.1 Introduction and historical aspects

Scientific and public interest in water quality is not new. In the United Kingdom (UK), it probably had its origins in the mid-eighteenth century. In 1828, the editor of Hansard, Mr. John Wright, published anonymously a pamphlet attacking the drinking water quality in London. This led to the setting up of a Royal Commission, which established the principle that water for human consumption should at all times be 'wholesome'. The term wholesome has been incorporated into virtually every piece of legislation concerned with drinking water ever since.

The first unequivocal demonstration of water-borne transmission of the cholera disease was by Snow in 1854. This stimulated great advances in water treatment practices, in particular the routine application of slow sand filtration and disinfection of public water supplies.

Although the Royal Commission of 1828 was concerned with water quality, it had difficulty in defining it precisely, there being virtually no analytical techniques available at the time with which to determine either microbial or chemical contamination. Consequently, since that time, there has been a continuing and often fierce debate on what constitutes a suitable quality for human drinking water. Not surprisingly, in the nineteenth and early part of the twentieth century, the evaluation was largely based on subjective, usually sensory, perception.

Many authorities (for example, Sir Edwin Chadwick) believed an atmospheric 'miasma' above the water, rather than the water itself, was responsible for disease transmission. As a consequence, great efforts were made to remove the smell, assuming that this would dispel the disease. In 1856, during the 'great stink', sheets drenched in chemicals were hung from the windows of the Houses of Parliament to exclude the smell. This action did at least focus the minds of the politicians on the need to take action to improve the quality of London's water supply.

Even today, taste, smell and appearance (colour and turbidity) are considered useful criteria for judging water quality. However, in addition, there are now objective methods for determining the presence and level of many (but by no means all) of the microbial contaminants likely to be present in drinking water.

Since the 1960s, the emphasis regarding drinking water quality has shifted from the bacteriological quality to the identification of chemical contaminants. This reflects largely the very considerable success of the water industry in overcoming bacteriological problems, although this victory is not complete; i.e. many viruses and *Cryptosporidium* cause public health concerns.

With the great methodological improvements in analytical chemistry over the past 50 years, it was recognised that water contains trace amounts of several thousand chemicals and that only the limitations of analytical techniques restrict the number of chemicals that can be identified. Many of these chemicals are of natural origin, but pesticides, human and veterinary drugs, industrial and domestic chemicals, and various products arising from the transport and treatment of water are very commonly found, albeit normally at very low concentrations.

In addressing the problem of the contribution of water-borne chemicals to the incidence of human disease, water scientists, whose previous experience has typically been confined to microbiological problems, have tended to focus on acute risks. The absence of detectable short-term adverse effects of drinking water has been taken by many as conclusive evidence that the presence of such chemicals is without risk to man.

While information on the acute toxicity of a chemical can be very useful in determining the response to an emergency situation such as an accidental spillage or deliberate release of chemicals into a watercourse or even into the water supply, such information is of little use in predicting the effects of daily exposure to a chemical over many years.

However, low levels of chemicals are much more likely to cause chronic than acute effects to health. Here, direct reliable information is very sparse. Some authorities appear to have accepted the 'naïve' assumption that information on the acute effects of a chemical, in either man or experimental animals, can be used to predict the effects of being exposed over a lifetime. In practice, the chronic effects of a chemical have rarely any resemblance to the acute effects.

An evaluation of health risks associated with drinking water is necessary and timely. If we are to obtain a proper assessment of the health risk that could arise in humans through exposure to chemicals in water over a lifetime, understanding must be developed on the following:

- Identification of the chemicals, which are of most concern;
- Data on the effects of long-term exposure in humans and/or animals to each chemical;
- A measure of the extent and form of exposure to each chemical;

- Identification of particularly at-risk groups; and
- The means of establishing how exposure to other chemicals in the water can modify the toxicity.

1.2 Water quality standards and treatment objectives

It is commonly agreed that there are three basic objectives of water treatment, namely:

(1) Production of water that is safe for human consumption;
(2) Production of water that is appealing to the customer; and
(3) Production of water treatment facilities, which can be constructed and operated at a reasonable cost.

The first of these objectives implies that the water is biologically safe for human consumption. It has already been shown how difficult it is to determine what 'safe' actually means in practice. A properly designed plant is not a guarantee of safety, standards will change and plant management must be flexible to ensure continued compliance.

The second basic objective of water treatment is the production of water that is appealing to the customer. Ideally, an appealing water is one that is clear and colourless, pleasant to taste, odourless and cool. It should be non-staining, non-corrosive, non-scale forming and reasonably soft. The consumer is principally interested in the quality of the water delivered to the tap, not the quality at the treatment plant. Therefore, storage and distribution need to be accomplished without affecting the quality of the water; i.e. distribution systems should be designed and operated to prevent biological growths, corrosion and contamination.

The third basic objective of water treatment is that it can be accomplished using facilities with reasonable capital and operating costs. Various alternatives in plant design should be evaluated for cost-effectiveness and water quality produced.

The objectives outlined need to be converted into standards so that proper quality control measures can be used. There are various drinking water standards. The key variables are as follows:

- Organoleptic parameters: colour, turbidity, odour and taste.
- Physical-chemical parameters: temperature, pH, conductivity, dissolved oxygen, dissolved solids, chlorides, sulphate, aluminium, potassium, silica, calcium, magnesium, sodium, alkalinity, hardness, and free carbon dioxide (CO_2).
- Parameters concerning undesirable substances: nitrate, ammonium, total organic carbon (TOC), hydrogen sulphide, phenols, dissolved hydrocarbons,

iron, manganese, suspended solids, and chlorinated organic compounds other than pesticides.

- Parameters concerning toxic substances such as arsenic, mercury, lead and pesticides.
- Microbiological parameters: total coliforms, faecal coliforms, faecal streptococci, sulphite reducing clostridium and total bacterial count.

Standards usually give two values: a guide level (GL) and a maximum admissible concentration (MAC). The GL is the value that is considered satisfactory and constitutes a target value. The MAC is the value that the corresponding concentration in the distributed water must not exceed. Treatment must be provided, where the concentration in the raw water exceeds the MAC.

Standards also specify the methods, frequencies and nature of the analysis. For total hardness and alkalinity, the standards specify minimum values to be respected when water undergoes softening.

Most standards group substances into five categories, namely:

(1) Microbiological;
(2) Inorganic with consequences on health;
(3) Organic with consequences on health;
(4) Appearance; and
(5) Radioactive components.

1.3 Some thoughts on the standards

One of the main sources of confusion regarding water standards and their interpretation is the lack of any clear indication as to how the standard was derived. This results in the interpretation of all standards as 'health standards' by the public, and subsequently in the difficulty of assessing what should be done by the water supplier, if a threshold is exceeded.

This is particularly true of drinking water quality directives as insufficient explanation of the derivation of the actual numbers is often given. There are even thresholds for variables regarded as toxic, which are based on political or other considerations, and are therefore only loosely based on science (e.g., pesticides). The use of such approaches is acceptable as long as the reasoning behind them is clear to all.

International guidelines are usually intended to enable governments to use them as a basis for standards, taking into account local conditions. They are intended to be protective of public health and they should be absolutely clear, even down to the detailed scientific considerations such as derivation of uncertainty factors and the rounding of numbers. It is therefore incumbent on the expert groups to justify their thinking and present it openly for all to see. Such a discipline avoids the 'fudging' of issues while giving the impression of scientific precision, and can only be of value in increasing public confidence in the resulting guidelines.

It is clear that at present, standards for water quality are as follows:

- Loosely based on science (though the situation is improving);
- Not static (the science of monitoring as well as our understanding of the health implications of chronic exposure of many contaminants is improving); and
- Important in the quality control of potable water (both for supplier and consumer).

Concerning the outflow water quality of most wetland systems, standards are either unclear or are currently being developed. The local environment regulator usually sets standards for specific wetland system applications.

Chapter 2

Water treatment

2.1 Sources of water

The source of raw water has an enormous influence on the water chemistry and consequently its treatment. Raw water is commonly abstracted from one of the four sources:

1. Boreholes extracting groundwater: This water is usually bacteriologically safe as well as being aesthetically acceptable. It may require some treatment such as aeration or softening.
2. Rivers: Water can be abstracted at any point along the length of a river; however, the further downstream, the more likely the water to require considerable treatment.
3. Natural lakes: The degree of treatment required for lake water depends on a number of factors such as the catchment use in the immediate vicinity of the lake, the lake trophic status, and the presence of sewage treatment works.
4. Man-made lakes and reservoirs: These are similar to lakes and rivers, but better managed. The degree of treatment depends on the management of the catchment and upstream catchment usage.

Water for domestic consumption may also come from other sources such as sea water (desalination) or treated sewage effluents.

2.2 Standard water treatment

The purpose of screens is simply to remove solid floating objects, which may cause damage or blockage in the plant in the raw water; e.g., logs and twigs. Sometimes a much finer screening is carried out, called straining. This is usually performed on lake and reservoir water to remove algae.

A coagulant is added to the raw water to de-stabilise the colloidal material in the water (Chapter 9). Commonly used chemicals are:

- Alum (aluminium sulphate) $Al_2(SO_4)_3 \cdot nH_2O$;
- Ferric chloride $FeCl_3$;
- Ferrous sulphate (copperas) $FeSO_4 \cdot 7H_2O$;
- Lime (burnt CaO; slaked $Ca(OH)_2$); and
- Polyelectrolytes (long chain organic molecules normally used in conjunction with a conventional coagulant).

In order for the coagulant to function efficiently, it must be rapidly and uniformly mixed through the raw water. This usually takes place in a high shear (turbulent) environment such as one induced by a hydraulic jump (low cost option), pump, jet mixer or propeller mixer.

After the coagulant is uniformly distributed in the water, it requires time to react with the colloid, and then further time (and gentle agitation) to promote the growth (agglomeration) of settleable material (flocs). This is generally accomplished in either a tank with paddles (mechanical mixing) or through a serpentine baffled tank (hydraulic mixing). Once flows of a settleable size have formed, they are removed usually by sedimentation (sometimes by floatation).

In the UK and Ireland, developments in the 1940s led to the introduction of the sludge blanket clarifier (Chapter 10). This is a single unit which encompasses rapid mixing, flocculation and settling.

In order to remove either solids carried over from settling tanks and/or any uncoagulated material (organic or inorganic), a sand bed filter is provided. The water flows downwards through the bed and the impurities are removed by attachment to the sand grains. The sand grains therefore require periodic cleaning. The frequency of cleaning depends on the type of filter used. The two commonly used types are the following (Chapters 12 and 13):

- Rapid gravity sand filter (high loading rate); and
- Slow sand filter (loading rate is approximately one-tenth of that for rapid gravity sand filter).

Sometimes fluoride is added to the water to reduce the incidence of dental caries. This is a process which provokes public debate. In the USA and UK, chlorine is usually added to the water to disinfect it (Chapter 22). It follows that the water is bacteriologically safe when it leaves the treatment works, and excess chlorine is added to protect the water from contamination during the distribution process.

There are several other commonly used processes. Their usage depends on the nature of the raw water. Air can be introduced to the water to oxidise impurities; e.g., iron, manganese or chemical compounds affecting the taste of water. pH control is a common process since many of the chemical treatment processes are pH-dependant. Softening reduces the hardness and/or alkalinity of water to improve its aesthetic acceptability. This is a complex chemical process depending on the nature of both the anion (HCO_3^-, CO_3^{2-} or OH^-) and the cation (Ca^{2+} or Mg^{2+}).

2.3 Basic water chemistry

The most important chemical variables of raw water are usually taken as pH and alkalinity. Alkalinity consists of those chemical species that can neutralise acid. In other words, these species allow the water to resist changes and provide buffering capacity. The major constituents of alkalinity are the hydroxyl (OH^-), carbonate (CO_3^{2-}) and bicarbonate (HCO_3^-) ions. The relative quantities of each are a function of pH.

No significant concentration of hydroxyl ions exists below pH 10, and no significant carbonate concentration can be detected below pH 8.5. For most waters, alkalinity thus consists of the bicarbonate ion. The other species may be formed in the treatment process. The bicarbonate and carbonate ions in the water result from the dissolution of carbonate rocks.

The pH is a measure of the free hydrogen ion concentration in water. Water, and other chemicals in solution, will ionise to a greater or lesser degree. The ionisation reaction is given in Eq. (2.3.1).

$$H_2O \Leftrightarrow H^+ + OH^- \qquad (2.3.1)$$

In neutral solutions, the [OH^-] activity is equal to the [H^+] activity. Hence the pH and pOH are both equal, and have the numerical value of 7. An increase in acidity, for example, leads to higher values of [H^+], thus lowering the pH.

The various chemical reactions, which occur in natural waters and in process water, are generally considered to occur in dilute solutions. This permits the use of simplified equilibrium equations in which molar concentrations are considered to be equal to chemical activities.

The assumption of dilute conditions is not always justified, but the error introduced by the simplification is no greater than the error, which might be introduced by competing reactions with species that are not normally measured in water treatment.

Concentrations of different chemical species in water may be expressed in moles per litre, in equivalents per litre or in mass per unit volume (typically mg/l). The equivalent of a species is its molecular weight divided by the net valence or by the net change in valence in the case of oxidation and reduction reactions.

The number of equivalents per litre (normality) is the concentration divided by the equivalent weight. The number of moles per litre is called the molarity.

Chapter 3

Sewage treatment

3.1 Introduction

The waste disposed of by domestic households and industry is conveyed to the treatment works by means of pipes (sewers). The arrangement of sewers is known as the sewerage system. Everything that flows in the sewers is sewage. These terms are often confused in practice.

In a traditionally combined sewer, all sewage, both foul and surface water, is conveyed in a single pipe. A foul sewer conveys the 'nasties'; i.e. contaminated water. A surface water sewer conveys the runoff from roofs and paved areas.

Concerning separate systems, two pipes are laid in the trench for the sewerage system; one for the foul sewer, and the second for the surface water. This book is predominantly concerned with the treatment of urban runoff.

The flow in a sewer can be estimated with Eq. (3.1.1). The mean domestic water consumption is typically 140 l/h/d for rural and 230 l/h/d for urban areas.

$$DWF = PQ + I + E \qquad (3.1.1)$$

where,

DWF = averaged total flow in 24 h (dry weather flow) $\left(\frac{Q_T}{24}\right)$;

P   = population;
Q   = mean domestic water consumption;
I   = rate of infiltration;
E   = industrial effluent discharge to the pipe; and
Q_T = total volume of flow in a 24-h period.

3.2 Design flow rates

Normally, at sewage treatment works, flows up to 3 dry weather flow (DWF) are given full treatment; >6 DWF (since they are diluted by the surface water) require only preliminary treatment. Flows between 3 and 6 DWF are stored temporarily and given full treatment.

However, care needs to be taken in the design of overflow structures, particularly for flows >6 DWF. These must be designed such that the outflow from them has a minimum impact on the receiving water; in particular, care must be taken with the solid material, which occurs in the so-called first 'foul flush'; i.e. immediately after the rainfall storm commences, accumulated material in the sewer is likely to be flushed out of the system.

3.3 Treatment principles

Typically, raw sewage contains 99.9% water and 0.1% solids. The sewage treatment process is fundamentally about separating solids from the water. The treatment of solids and sludge forms an important and costly area of sewage treatment. The impurities in the sewage can be categorised as follows:

- Floating or suspended solids (e.g., paper, rags, grit and faecal solids);
- Colloidal solids (e.g., organics and micro-organisms);
- Dissolved solids (e.g., organics and inorganic salts); and
- Dissolved gases (e.g., hydrogen sulphide and carbon dioxide).

These impurities are removed from the sewage using operations or processes, which are either physical, chemical or biological in nature. Physical operations depend on the physical properties of the impurity for efficient removal; e.g., screening, filtration and sedimentation. Chemical operations depend on the chemical properties of the impurity and those that utilise the chemical properties of additives for efficient removal; e.g., coagulation, precipitation and ion exchange. Biological processes utilise biochemical and/or biological reactions to remove soluble or colloidal organic impurities; e.g., percolating filters and activated sludge.

3.4 Engineering classification of sewage treatment stages

Wastewater engineers tend to describe the sewage treatment process in terms of the stages of treatment as shown below:

- Preliminary treatment (physical); e.g., screening and grit removal;
- Primary treatment (physical and/or chemical); e.g., sedimentation and flotation;

- Secondary treatment (biological and/or chemical); e.g., constructed wetlands, biological filters and the activated sludge process; and
- Tertiary treatment (physical and/or chemical and/or biological); e.g., polishing wetlands, micro-straining, grass plots and lime precipitation).

At the secondary treatment stage, either percolating filters or activated sludge treatment is usually present, certainly not both in parallel. On occasions, when treating industrial wastes, they may both be used, but always in series. It should be noted that sludge is produced at the majority of the treatment stages. However, in normal practice, the works are organised such that all sludge is collected centrally.

Wetland systems can be designed for each engineering stage, and for sludge treatment. However, constructed treatment wetlands are usually applied for secondary or tertiary treatment stages. Wetlands integrated in sustainable urban drainage systems (SUDS) are frequently used for preliminary and primary treatment purposes. Urban runoff requires full treatment, which is usually not the case in practice, unless for combined sewer systems and minor storms.

Chapter 4

Organic effluent

4.1 Biochemical oxygen demand

When wastewater including urban runoff is discharged into a watercourse, it exerts a polluting load on that water body. Micro-organisms present in the natural water and the wastewater break down (stabilise) the organic matter. In permitting discharges to watercourses, the Environment Agency in England, for example, tries to ensure that the conditions are aerobic, so that all other life forms in the river (e.g., fish) can continue to survive. The early forms of wastewater treatment developed are aerobic, and so the simplest way of estimating the biodegradability of a wastewater is to estimate the amount of oxygen required to stabilise the waste.

In order to devise an easy and simple method of assessing the oxygen demand, the following constituents of a closed system should be considered:

- Air (in excess);
- Small number of bacteria; and
- Finite amount of substrate (waste representing food).

The following phases of biological growth and decline can be identified in such a system:

- Lag phase: Bacteria are acclimatising to system conditions, and, in particular, the substrate; very little increase in numbers.
- Log growth: Bacteria are acclimatised; food is not a limiting factor; rapidly increasing population of bacteria.
- Declining growth: Food eventually becomes limiting; declining growth rates.
- Endogenous respiration: As the substrate concentration becomes depleted, competition increases; bacteria start consuming dead bacterial cells, and eventually start consuming live cells.

It is a system of this type that is used to assess the oxygen demand of wastes including organic matter from urban runoff. The test developed from this system is the biochemical oxygen demand (BOD) test.

4.2 BOD test

The BOD test is carried out as follows: a known quantity of a wastewater sample (suitably diluted with a prepared water) is placed in a 300-ml BOD bottle. The prepared water is saturated with dissolved oxygen (DO), and nutrients and a buffer are added. The bottles are then sealed, so that they are airtight. The bottles are subsequently incubated at 20°C in the dark (Clesceri et al., 1998).

Initially, the bacteria break the carbon-based molecules down. In practice, a second oxygen demand is observed. In the case of raw sewage, this stage usually becomes apparent after approximately eight days of incubation at 20°C. This second stage is due to the oxidation of ammonia present in the waste; this is called nitrification. A large percentage of the nitrogen in the wastewater originates from proteins, the protein molecules are degraded to release ammonia. The oxidation process is described in Eqs (4.2.1) and (4.2.2)

$$2NH_4^+ + 3O_2 \xrightarrow{\textit{Nitrosomonas}} 2NO_2^- + 4H^+ + H_2O \tag{4.2.1}$$

$$2NO_2^- + O_2 \xrightarrow{\textit{Nitrobacter}} 2NO_3^- \tag{4.2.2}$$

Nitrification consumes a significant amount of oxygen, so that the total demand for nitrification is often comparable with the carbonaceous demand. Nitrification also generates protons (H^+ ions), which increase the acidity (pH) of the waste.

Traditionally, the BOD test is carried out for five days; the resulting oxygen demand is referred to as the BOD_5. The BOD is calculated as follows (Eqs (4.2.3) and (4.2.4)):

$$BOD\ (mg/l) = \frac{\text{Initial DO in bottle} - \text{Final DO in bottle}}{\text{Dilution ratio}} \tag{4.2.3}$$

where,

$$\text{Dilution ratio} = \frac{\text{Volume of wastewater}}{\text{Volume of BOD bottle}} \tag{4.2.4}$$

In practice, the test is often modified slightly in that a quantity of seed micro-organisms are added to the BOD bottle to overcome the initial lag period. In this variant, the BOD is calculated from Eq. (4.2.5):

$$BOD = \frac{(D_1 - D_2) - f(B_1 - B_2)}{DR} \qquad (4.2.5)$$

where,

D_1 = dissolved oxygen initially in seed and waste bottle;
D_2 = dissolved oxygen at time T in seed and waste bottle;
B_1 = dissolved oxygen initially in seed only bottle;
B_2 = dissolved oxygen at time T in seed only bottle;
f = ratio of seed volume in seeded wastewater to seed volume in the BOD test on seed only; and
DR = dilution ratio.

Additional bottles are incubated. These contain only seed micro-organisms and dilution water to get the BOD of the seed, which is then removed from the BOD obtained for waste and seed.

However, the BOD test has two major disadvantages; it takes five days to obtain the standard test result, and the results can be affected by the process of nitrification (see above). Therefore, a nitrification inhibitior is often used (Chapter 26).

4.3 Chemical oxygen demand

The disadvantages of the BOD test have led to the development of a simpler and quicker test. This test is known as the chemical oxygen demand (COD) methodology. In this test, strong chemical reagents are used to oxidise the waste. Potassium dichromate is used in conjunction with boiling concentrated sulphuric acid and a silver catalyst. The waste is refluxed in this mixture for two hours. The consumption of the chemical oxidant can be related to a corresponding oxygen demand (Clesceri et al., 1998).

The COD test oxidises material that micro-organisms cannot metabolise in five days or that are toxic. If the COD $\gg$ BOD in raw wastewater, then the waste is not readily biodegradable, and may be toxic to the micro-organism. If COD $\approx$ BOD, then the waste is readily biodegradable.

4.4 Other variables used for the characterisation of wastewater

Most wastewater treatment processes operate best in pH ranges between 6.8 and 7.4; indeed pH > 10 is likely to kill large numbers of bacteria. Suspended solids (SS) is a measure of the total particulate matter content of wastewater. The nature of the SS is likely to vary considerably depending on the nature of the waste.

The two most important nutrients in wastewater treatment are nitrogen and phosphorus, both are needed for cell growth. Nitrogen (N) is used in the protein synthesis (e.g., new cell growth). Phosphorus (P) is used for cell energy storage, and is usually present as orthophosphate (PO_4).

Organic nitrogen is associated with cell detritus and volatile SS. Free ammoniacal nitrogen (NH_3–N) results from the decay of organic nitrogen. Nitrite–nitrogen (NO_2–N) is formed in the first step in nitrification. Nitrate–nitrogen (NO_3–N) results from the second and final stage in the nitrification process.

For proper micro-organism growth, the ratio of C:N:P is important. Carbon (C) is measured by BOD_5. Nitrogen is measured by organic nitrogen and NH_3–N. However, NO_3–N is difficult for micro-organisms to use in their growth process. Phosphorus is measured as acid hydrolysable ortho-phosphate (PO_4). To achieve growth, the required minimum values for the C:N:P relationship are 100:5:1.

Chapter 5

Stream pollution and effluent standards

5.1 Organic stream pollution

Since most effluents including storm runoff are discharged to a receiving water-course, it is important that the concentration of the effluent is such that the receiving water can assimilate the waste and break it down. The receiving water should also remain in a condition appropriate to its use (usually aerobic).

Consider the discharge of an organic effluent into a river. In the receiving water, there are two processes taking place: oxidation of the organic waste and reaeration, and the introduction of oxygen into the water. Before the effluent is discharged, the river contains dissolved oxygen (DO). The effluent reduces the initial DO concentration progressively to satisfy the BOD.

In the zone of degradation, reaeration is smaller than the rate of decomposition. In this section, the decomposition of the effluent dominates, and so the DO concentration drops rapidly. The sediment accumulation in the immediate vicinity may be large due to the settling of suspended material in the effluent.

In the zone of active decomposition, reaeration is approximately equal to the rate of decomposition. The water is likely to contain little diversity of life forms; the bottom sediments may possibly be anaerobic.

In the zone of recovery, the rate of reaeration is larger than the rate of decomposition. Since the effluent oxygen demand is dropping and the DO deficit is large, then atmospheric oxygen will diffuse into the water body at a greater rate, thus the DO begins to increase. Nitrification is also likely to begin to take place, and the life forms present in the river increase in diversity.

In the zone of clear water, the DO has now returned to its original value, the BOD has been virtually eliminated, only a background level remains. However, the river has been permanently changed; there are increased levels of nutrients in the water, and this may lead eventually to eutrophication.

5.2 Prediction of organic stream pollution

In the absence of reaeration, Eq. (5.2.1) should be considered:

$$\frac{dD}{dt} = K_1 L_t \qquad (5.2.1)$$

where,

D = dissolved oxygen deficit (DOsat − DOactual);
K_1 = reaction rate coefficient; and
L_t = ultimate BOD remaining at time T.

In water bodies, reaeration usually occurs as a result of a difference in partial pressures and the turbulence in the river flow. Reaeration can be expressed with Eq. (5.2.2):

$$\frac{dD}{dt} = -K_2 D \qquad (5.2.2)$$

where,

K_2 = reaeration coefficient.

Note that the negative sign indicates that oxygen transfer tends to reduce the deficit. Combining both decomposition and reaeration, Eq. (5.2.3) can be obtained.

$$\frac{dD}{dt} = K_1 L_t - K_2 D \qquad (5.2.3)$$

This is a simple differential equation, which can be easily integrated (with appropriate limits) to yield the Streeter–Phelps equation (Eq. (5.2.4)):

$$D_t = \frac{k_1 L_0}{k_2 - k_1} \left[10^{-k_1 t} - 10^{-k_2 t} \right] + D_0 10^{-k_2 t} \qquad (5.2.4)$$

where,

D_0 = dissolved oxygen deficit at time $t = 0$;
L_0 = ultimate BOD initially (i.e. $t = 0$);
L_t = $L_0 \, 10^{-k_1 t}$;
k_1 and k_2 = reaction rate and reaeration coefficient (base 10); and
D_t = dissolved oxygen deficit at time t.

The critical point on the DO sag curve is the minimum value of DO. The occurrence of the critical point can be defined with Eqs. (5.2.5)–(5.2.7).

$$k_1 L - k_2 D_c = 0 = \frac{dD}{dt} \tag{5.2.5}$$

$$\therefore D_c = \frac{k_1}{k_2} L_0 10^{-k_1 t_c} \tag{5.2.6}$$

$$t_c = \frac{1}{k_2 - k_1} \log_{10} \left[\frac{k_2}{k_1} \left(1 - \frac{D_0 (k_2 - k_1)}{L_0 k_1} \right) \right] \tag{5.2.7}$$

where:

D_c = critical deficit reached at time t_c.

The Streeter–Phelps equation is only valid when no change in the pollution load or dilution occurs. Complex discharge and river problems require to be solved by a stepwise process.

5.3 Effluent discharge standard principles

Water quality standards should achieve the following:

• Safeguard public health;
• Protect water, so that it is suitable for abstraction, and subsequent use in domestic, agricultural or industrial circumstances;
• Cater for the needs of commercial, game and course fisheries; and
• Cater for the relevant water-based amenities and recreational requirements.

The limits or standards placed on effluent discharges have been traditionally specified in terms of effluent volume, BOD and SS. Standards are based traditionally on the fact that a pollution-free stream would have a $BOD_5 = 2$ mg/l, and if the $BOD_5 > 4$ mg/l, the stream may become a nuisance; i.e. occasionally anaerobic. However, these standards are now superseded by various national standards and international directives. Nevertheless, standards for urban water runoff are currently being developed.

Chapter 6

Preliminary treatment

6.1 Introduction

Screening is normally the first operation used at sewage treatment plants, and just before urban runoff is conveyed into a wetland system. The general purpose of screens is to remove large objects such as twigs, rags, paper, plastic, metals and shopping trolleys. These objects, if not removed, may damage the pumping and sludge removal equipment, hang over weirs, and block valves, nozzles, channels and pipelines, thus creating serious plant operation and maintenance problems.

The second operation constituting preliminary treatment is grit removal. Grit includes sand, dust and cinders. These are non-putrescible materials with a specific gravity greater than that of organic matter. It is necessary to remove these materials in order to:

- Protect moving mechanical equipment and pumps from unnecessary wear and abrasion;
- Prevent clogging in pipes and heavy deposits in channels;
- Prevent cementing effects on the bottom of sludge digesters and primary sedimentation tanks; and
- Reduce accumulation of inert material in aeration basins and sludge digesters, which would result in the loss of usable volume.

6.2 Design of screening units

For preliminary treatment, screens normally comprise vertical or inclined bars with openings between 20 and 60 mm for coarse screens, and 10 and 20 mm for medium screens. The bars are usually made from steel, and are between 50 and 75 mm wide, and between 10 and 15 mm thick. The spacings used are normally approximately 20 mm for mechanically raked and between 25 and 40 mm for manually cleaned screens.

The hydraulic design of screening units must include the calculation of screen area, and the head loss through the screen. The area of submerged screen surface is based on the velocity of flow through the clean openings. The corresponding velocity is 600 mm/s for average flows, and 900 mm/s for the maximum rate of flow.

The largest area is the controlling area. For small and manually cleaned units, the minimum working width is 450 mm. However, the usual width is between 600 and 900 mm for mechanically cleaned units. In general, the effective screen area should be about two times the cross-sectional area of the incoming pipe such as a sewer, for example. The head loss through the bars is calculated using Eqs. (6.2.1)–(6.2.3).

$$h_L = \frac{V^2 - V_5^2}{2g} \cdot \frac{1}{0.7} \qquad \text{(clean or partly clogged bars)} \tag{6.2.1}$$

$$h_L = \beta \left(\frac{W}{b}\right)^{\frac{4}{3}} h_v \sin\theta \qquad \text{(clean screens)} \tag{6.2.2}$$

$$h_L = \frac{1}{2g} \left(\frac{Q}{CA}\right)^2 \qquad \text{(common orifice equation used for fine screens)} \tag{6.2.3}$$

where,

h_L = head loss through bars (m);
V = velocity through bars and in the channel upstream of unit (m/s);
W = maximum cross-sectional width of bars facing the direction of flow (m);
b = minimum clear spacing of bars (m);
h_v = velocity head of flow approaching the bars (m);
θ = angle of bars with horizontal (°);
A = effective submerged open area (m^2);
C = coefficient of discharge (i.e. 0.6 for a clean screen); and
β = bar shape factor with different β values for different shapes (sharp-edged rectangular, 2.42; rectangular with semi-circular upstream face, 1.83; circular, 1.79; rectangular with semi-circular faces upstream and downstream, 1.67; 'tear drop' shape, 0.76).

The width of the screen channel is usually calculated with Eq. (6.2.4).

$$W = \frac{B + b}{b} \cdot \frac{Q}{VD} \tag{6.2.4}$$

where,

W = width of the channel excluding supports (m);
B = width of screen bars (m);
b = spacing of screen bars (m);
Q = maximum flow rate (m^3/s);
V = maximum velocity of flow through screens (m/s); and
D = depth of flow at screens (m).

Screens can either be manually cleaned (not popular but common for wetland systems) or mechanically cleaned. In mechanically cleaned screens, cleaning is either continuous, or initiated when the differential head through the screen reaches 150 mm.

6.3 Design details for screening units

Good access is required, because screening units need frequent inspection and maintenance. Ample working space must be provided for screens in deep channels. Protective structures such as guards are traditionally omitted, though trends are changing. Good corrosion protection must be written into specifications.

Cleaning mechanisms such as front- or back cleaning with front- or back discharge are usually designed by the manufacturer. Back-cleaning devices are protected from damage by the screen itself. Front discharge by macerated screenings is preferable to back discharge.

The following types of operating controls are used for mechanically cleaned screens, sometimes in combination: manual stop-start, automatic stop-start by time control, high-level alarm, differential head actuated automatic stop-start, and overload switch and alarm.

6.4 Comminutors

Comminutors are combined screen and macerator units. They consist of an electrically rotated drum with horizontal slots that form a screen. The sewage gravitates from the upstream channel into a spiral flow channel, through the slots and open bottom of the drum, and into the downstream channel via an inverted siphon. Suspended solids and floating solids are held by the liquid flow against the outside of the drum, and are macerated by stationary cutters.

It is an advantage of comminutors that screenings do not have to be removed from the flow. A disadvantage is the tendency for 'stringing' or 'balling up' of material and the head loss is higher than that with screens.

7.9.2.3 Imhoff flow tanks

Imhoff tanks are designed to provide sedimentation and digestion volume. They may be either round or rectangular (usually the latter). Either design requires deep excavation and a somewhat more costly construction than either horizontal-flow or radial-flow tanks.

Sewage enters these tanks below the top water level, but above the maximum sludge level. The flow direction is upwards while settled sludge sinks to the bottom of the tank. The allowable upward velocity determines the surface area.

The side slopes of the cone should be >60° to the horizontal. The inlet and outlet should be designed to produce low velocities. The use is normally confined to small works. The sludge can be drawn off hydrostatically approximately once a week under manual control. The capacity provided within the tanks (excluding sludge capacity) must be between 5 and 15 h of a DWF.

7.9.2.4 Radial-flow tanks

Continuous flow tanks have detention periods at a DWF of between 6 and 10 h. The design is based on the criteria presented above. The effect of the scraper support on the time ratio must be considered (in model tests) after initial sizing of the units. Floor slopes vary between 0° (flat) and 15° (commonly between 7.5 and 10°). V-notch weirs produce a more even flow distribution at low flows, but tend to reduce the efficiency slightly.

Automatic descumming is carried out by the rotating scraper; scum boxes being provided on the tank periphery. Sludge is deflected into a central hopper, having a capacity of one day. The designs of scrapers are many and varied. All incorporate renewable rubber tips.

Radial flow tanks cannot be built in batteries, but can be cheaper than rectangular tanks since they are shallower and have less complicated sludge scrapers. If the same tank diameter can be used for final or humus settlement, re-usable formwork may further cut costs.

7.9.3 Secondary sedimentation

Concerning secondary sedimentation tanks, the SS tend to be smaller and less dense having a lower settling velocity in comparison to the SS in primary sedimentation tanks. Wind can affect the settlement of sludge in secondary tanks, especially, if they are large and shallow. The capacity is normally fixed at a DWF between 4 and 6 h. The outlet channels are usually lined with white tiles to assist the visual inspection of the effluent. There are two classes of secondary sedimentation tanks depending on the biological treatment process.

Humus tanks for filtration plants are normally radial-flow tanks, although other types are possible. Sludge removal is usually intermittent. Humus sludge is returned to the feed channel of the primary sedimentation tank, and settled with the primary solids, since it assists sedimentation and is easier to dewater when mixed with primary sludge.

Concerning final tanks for activated sludge plants, sludge removal must be continuous to feed the aeration units with fresh sludge. Surplus activated sludge is returned to the inlet of the primary tank for the purpose of settlement.

Circular settlement tanks are usually similar to those used for primary and humus sedimentation. Normally, they are flat-bottomed with continuous removal of sludge by air lift or siphon draw off.

7.10 Sedimentation aids

Sewage includes colloidal particles, which may remain suspended indefinitely. However, these very finely divided particles tend to flocculate with the aid of mechanical agitation, aeration or chemical coagulation. The quality of the final effluent can thus be significantly increased.

Mechanical flocculation is normally achieved in double-zone tanks incorporating an inner mixing zone (paddles rotating at $\leq$450 mm/s), and another conventional settlement zone; e.g., Dorr Clariflocculator, which gives a 20% better effluent than that obtained by plain sedimentation at very little extra cost.

Chemical coagulants in common use are as follows:

- Hydrated lime: $Ca(OH)_2$.
- Aluminium sulphate: $Al_2(SO_4)_3$.
- Copperas: $FeSO_4$.

However, the misuse of chemicals can sterilise sewage or at least slow down the biological degradation rate. Sedimentation aids are usually unnecessary for most wetland systems due to the presence of plants that slow down the flow velocity of the storm runoff, for example.

Chapter 8

Theory of settling

8.1 Introduction

In most wastewater treatment works and wetland systems, sedimentation is the main process used in primary treatment. Primary sedimentation removes between 50 and 70% of the suspended solids (SS), containing between 25 and 40% of the BOD from the wastewater including urban runoff.

Sedimentation is normally defined as the removal of solid particles from a suspension by settling under gravity. Clarification is a similar term, which refers specifically to the function of a sedimentation tank in removing suspended matter from the water to give a clarified effluent.

Thickening occurs in sedimentation tanks ('clarifiers' in USA) and is the process whereby the settled impurities are concentrated and compacted on the floor of the tank and in the sludge hoppers. The concentrated impurities removed from the bottom of the tank are called sludge, the impurities floating to the surface of the tanks are called scum.

8.2 Classification of settling behaviour

8.2.1 Class I settling

The hydrodynamic problem of settling has been studied by many researchers, and a commonly accepted classification scheme is as follows:

- Class I: unhindered settling of discrete particles;
- Class II: settling of a dilute suspension of flocculent particles;
- Class III: hindered and zone settling; and
- Class IV: compression settling (compaction and consolidation).

Concerning Class I settling, a single particle in a liquid of a lower density should be considered for the purpose of this explanation. The particle will accelerate until a settling (or terminal or limiting) velocity is reached at the point where the gravitational force is balanced by the frictional drag force. The settling velocity can be determined using either Stoke's law (Eq. (8.2.1)) or Newton's law (Eq. (8.2.2)), depending on whether the flow past the particle is laminar or turbulent, respectively.

$$U_s = \frac{(S_s - 1)gd^2}{18v} \tag{8.2.1}$$

where,

U_s = settling velocity;
S_s = specific gravity of the solid matter;
g = gravitational constant;
d = diameter of the solid matter; and
v = kinematic viscosity.

$$U_s = \sqrt{3.3(S_s - 1)gd} \tag{8.2.2}$$

where,

U_s = settling velocity;
S_s = specific gravity of solid matter;
g = gravitational constant; and
d = diameter of the solid matter.

8.2.2 Class II settling

In many practical cases, even in a relatively dilute suspension, particles coalesce to form particle aggregates (i.e. flocculation resulting in flocs) having increased settling velocities. The extent of such flocculation is a function of many variables including suspension type and concentration, the prevailing velocity gradient, and time. In wastewater engineering, most sewage including urban runoff exhibits Class II (see also above) settling in primary tanks.

The expected removal of a flocculent suspension in a sedimentation process can be estimated from a laboratory settling test using a water column height equal to that used in the process. At various time intervals, samples are withdrawn from ports at different depths and analysed for SS. Lines of equal percentage removal are then interpolated between the plotted points. The resulting curves can be used to determine the overall removal of solids for any detention time and depth within the range of the data, bearing in mind that the test conditions are quiescent.

8.2.3 Class III and Class IV settling

Concerning Class III settling, as the concentration of particles in a suspension is increased, a point is reached where particles are so close together that they no longer settle separately, but the velocity fields of the fluid displaced by adjacent particles overlap. This gives rise to a net upward flow of liquid displaced by settling particles. This results in a reduction of the settling velocity, and is called hindered settling.

For traditional wastewater systems, Class III settling is most commonly exhibited in the sludge blanket clarifier, where the particle concentration is so high that the whole suspension tends to settle as a 'blanket' (termed zone settling). Furthermore, most urban runoff exhibits Class II (see above) or Class IV settling in a densely planted constructed treatment wetland.

At the bottom of a settling column, as the settling continues, a compressed layer of particles begins to form. The particles in this region apparently form a structure in which there is close physical contact between the particles. As the compression layer forms, regions containing successively lower concentrations of solids than those in the compression region extend upward in the cylinder.

Thus, the hindered-settling region contains a gradient in solids concentration increasing from that found at the interface of the settling region to that found in the compression zone. This region of Class IV settling is particularly important when considering the design of thickeners for activated sludge.

8.3 Ideal settling

The behaviour of an ideal sedimentation tank, operating on a continuous basis with a discrete suspension of particles can be studied as follows:

- Assume quiescent conditions in the settling zone;
- Assume uniform flow across the settling zone;
- Assume uniform solids concentration as the flow enters the settling zone; and
- Assume that solids entering the sludge zone are not resuspended.

Equation (8.3.3) shows how to calculate v, which is termed overflow rate, and which should be considered as a design settling velocity.

$$v = Q/A \qquad (8.3.3)$$

where,
v = overflow rate (m/s)
Q = disharge (m^3/s); and
A = tank surface area (m^2).

Chapter 9

Coagulation and flocculation

9.1 Introduction

When a material is truly dissolved in water, it is dispersed as either molecules or ions. The 'particle' sizes of a dissolved material are usually in the range between 2×10^{-4} and 10^{-3} μm. The 'particles' cannot settle and cannot be removed by ordinary filtration.

True colloidal suspensions and true solutions are readily distinguished, but there is no sharp line of demarcation. Colloidal particles are defined as those particles in the range between 10^{-3} and 1 μm. In urban water runoff, for example, colloidal solids usually consist of fine silts, clay, bacteria, colour-causing particles and viruses.

9.2 Colloidal suspensions

In water and wastewater treatment, there are generally two types of colloidal systems, both of which have water as the disperse phase: colloidal suspensions (i.e. solids suspended in water) and emulsions (i.e. insoluble liquids such as oils suspended in water). Features of a colloidal suspension are as follows:

- Colloids cannot be removed from a suspension by ordinary filtration. However, they can be removed by ultrafiltration or by dialysis through certain membranes.
- Colloidal particles are not visible under an ordinary microscope. However, they can be seen as specks of light with a microscope when a beam of light is passed through the suspension. This is caused by the Tyndall effect, which is the scattering of light by colloidal particles.
- Brownian motion prevents the settlement of particles under gravity. Some colloids can be removed by centrifugation.
- There is a natural tendency for colloids to coagulate and precipitate. Sometimes this tendency is countered either by mutual repulsion of the particles or by

the strong attraction between the particles and the medium in which they are dispersed (usually water). If these effects are strong and coagulation does not occur, the suspension is said to be stable.

When the main factor causing stability of a colloidal suspension is the attraction between particles and water, the colloids are said to be hydrophilic. When there is no great attraction between particles and water, and stability depends on mutual repulsion, the colloids are said to be hydrophobic. Hydrates of iron and aluminium form hydrophobic colloids in water; stability occurs through mutual electrostatic repulsion; i.e. like charges repel. Proteins, starches or fats form hydrophobic colloids in water. Stability can be achieved through the attraction between water and particles. Stable hydrophilic colloids are difficult to coagulate.

Colloids have a large surface area per unit volume; e.g., if a 10-mm cube were broken into cubical particles with dimensions of 10^{-2} μm, the surface area would be increased by a factor of 10^6 to some 600 m^2. Surface effects are therefore of significance; of these, the following two are important:

(1) The tendency for substances to concentrate on surfaces (adsorption); and
(2) The tendency for surfaces of substances in contact with water to acquire electrical charges (giving them electro-kinetic properties).

At the surface, electrical charge results either from the colloidal material's affinity, from some ions in the water or from the ionisation of some of the atoms (or groups of atoms), which leave the colloid. The surface charge attracts ions carrying a charge of opposite sign, and thus creates a cloud of 'counter ions', in which the concentration decreases as the distance from the particle increases.

As an example, clay particles in water are charged, because of isomorphous substitution, whereby certain cations of the crystal lattice forming the mineral may have been replaced by cations of a similar size, but lower charge. For example, Si^{4+} may be replaced by Al^{3+}, or Al^{3+} by Mg^{2+}. In both cases, the lattice is left with a residual negative charge, which must be balanced by the appropriate number of compensating cations in the dry clay. These may be fairly large ions such as Ca^{2+} or Na^{2+}, which cannot be accommodated in the lattice structure, so that these ions are mobile, and may diffuse into solution when the clay is immersed in water, leaving negatively charged clay particles.

For these and other reasons, colloidal particles in water are usually charged. The majority of particles encountered in natural waters are negatively charged. A suspension of colloidal particles as a whole has no net charge, since the surface charge of the particles is exactly balanced by an equivalent number of oppositely charged counter-ions in solution.

Furthermore, the distribution of these counter-ions is not random, since by electrostatic attraction they tend to cluster around charged particles. The combined system of the surface charge of a particle and the associated counter-ions in solution is known as an electrical double layer (stern layer).

In addition to the forces related to the electrical charges, colloidal particles, when close together, are subject to van der Waal's forces. These originate in the behaviour of electrons, which are part of the atomic or molecular system. The forces of attraction become significant only at short distances (e.g., ≤ 1 μm).

The source of energy may be either from Brownian motion or the relative movement in the water (fluid shear). If the energy of an impact is inadequate, the remaining repulsion will force the particles apart, forming a stable colloidal suspension.

9.3 Coagulation processes

To destabilise a hydrophobic colloidal suspension to permit coagulation to occur, it is necessary to reduce or eliminate the energy requirement. The van der Waal's forces cannot be manipulated, but the electrical forces can. The principle methods are as follows:

- Reduction or neutralisation of the charges on the colloid;
- Increase of the density of the counter-ion field, and thus reduction of the range of the repulsive effect (compaction of the double layer); and
- Permanent contact through molecular bridges between particles.

Neutralising charges on colloids may be accomplished by the addition of either multivalent ions or colloids (or both) having an opposite charge. These are frequently added as a chemical coagulant. The coagulating power of a chemical rises rapidly with its valence. For example, Al^{3+} and SO_4^{2-} ions are several hundred times more effective than Na^+ and Cl^- ions.

In water, which is almost pure (i.e. low concentration of ions), the cloud of counter-ions is widely dispersed. With increase in ionic strength of the solution, the counter-ion cloud becomes more concentrated near the colloidal particles, counter-acting the charge, and the field of influence of the colloidal charge becomes more restricted. Therefore, an increase in the ionic strength of a colloidal suspension will tend to cause destabilisation and consequently coagulation. This phenomenon has been observed in nature where turbid rivers run into brackish estuaries and marine wetlands, and coagulation and sedimentation subsequently occur.

9.4 Coagulation chemicals

Chemicals used for chemical coagulation of impurities in water treatment should be affordable and not leave any toxic or other undesirable residues in the water. Coagulation in water treatment occurs predominantly by two mechanisms:

- Adsorption of the soluble hydrolysis species on the colloid and destabili-sation; and

- Sweep coagulation where the colloid is entrapped within the precipitation product (e.g., aluminium hydroxide).

The specific mechanism occurring is dependent on both the turbidity and alkalinity of the water being treated. The reactions in adsorption and destabilisation are extremely fast and occur within 0.1 and 1 s. Sweep coagulation is slower, and occurs in the range between 3 and 17 s. The differences between the two types of coagulation in terms of rapid mixing are not delineated in the literature. It is imperative that the coagulants are dispersed in the raw water as rapidly as possible (≤ 0.1 s), so that the hydrolysis products that develop within 0.01 and 1 s will cause destabilisation of the colloid.

9.4.1 Aluminium compounds

The common aluminium salt used is aluminium sulphate (alum). It is available in solid, granular, powdered or liquid form. The chemical formula is $Al_2(SO_4)_3 \cdot nH_2O$, where n depends on the method of manufacture, but is usually in the range between 12 and 16.

When hydrolysed, alum produces sulphuric acid as well as the hydrate. For example, when forming the simplest hydrate $Al(OH)_3$, the hydrolysis reaction can be described with Eq. (9.4.1).

$$Al_2(SO_4)_3 + 6H_2O \; Al(OH_3) + 3H_2SO_4 \qquad (9.4.1)$$

Therefore, it is regarded as an acid salt and the water must contain enough alkalinity (natural or added) to react with the acid as it forms to maintain the pH within the desired range for good coagulation and flocculation. It can be shown that 1 mg/l alum is used for 0.5 mg/l alkalinity (as $CaCO_3$), which results in the production of 0.44 mg/l CO_2. After coagulation, taste requirements dictate a residual alkalinity of ≥ 30 mg/l.

9.4.2 Sodium aluminate

Sodium aluminate is an alkaline salt produced by treating aluminium oxide with caustic soda. The hydrolysis reaction is shown in Eq. (9.4.2).

$$NaAlO_2 + 2H_2O \rightarrow Al(OH)_3 + NaOH \qquad (9.4.2)$$

It can be used either in conjunction with alum or on its own for waters, which do not have sufficient natural alkalinity. However, subsequent coagulation may not be as good as when alum is used in combination with an alkali, because the divalent sulphate ions introduced with the alum have a favourable influence on coagulation.

9.4.3 Iron salts

Ferric ions can be hydrolysed and precipitated as ferric hydroxide at pH >4.5. Poor coagulation occurs in the pH range between 7 and 8.5. The ferric ion (like the aluminium ion) hydrolyses to form hydrates and an acid. Enough alkalinity must be present to combine with the acid and maintain a suitable pH for good coagulation. Unlike $Al(OH)_3$, ferric hydroxide does not redissolve in alkaline solutions, and so there is no particular upper pH limit for ferric coagulation. In addition, the ferric hydroxide floc is usually heavier, and hence faster settling, than the aluminium hydroxide floc.

Considering coagulation of coloured acid waters (typically of peat-based catchments), the presence of the divalent sulphate ion assists coagulation to a greater degree than that of the monovalent chloride ion. The commonest ferric salt used in water treatment is ferric chloride $(FeCl_3)$, which is very corrosive in the presence of water. Therefore, it requires transportation in rubber-lined tanks or glass containers.

Ferric sulphate $(Fe_2(SO_4)_3)$ is normally available as an anhydrous material, making transportation easier. On some occasions, ferrous salts are used. To be effective, the ferrous ion should be oxidised to the ferric form when in solution.

At pH values >8.5, dissolved oxygen (DO) in the water will cause oxidation. However, at lower pH values, chlorine can be used as an oxidising agent. Ferrous sulphate $(FeSO_4.7H_2O)$, also known as copperas, is usually available as a granular material. Because it requires a pH of ≥ 8.5 to be oxidised by DO, it cannot be used on its own in natural waters. When used in conjunction with lime, it is useful in coagulating the precipitate obtained in lime softening of water and removal of excessive iron and manganese from waters.

Chlorinated copperas can be used for water with a relatively low pH. It is formed by mixing solutions of copperas and chlorine (Eq. (9.4.3)).

$$6FeSO_4 + 3Cl_2 \rightarrow 2Fe_2(SO_4)_3 + 2FeCl_3 \qquad (9.4.3)$$

Moreover, an excess of chlorine can be used to disinfect the water. Copperas can be more readily used than the ferric salts, because it is not so corrosive and can be handled by ordinary chemical feeding equipment.

9.4.4 Coagulant aids

Lime is not normally used as a coagulant in water treatment, though it is commonly used in wastewater treatment. Lime is used in water treatment for adding alkalinity (pH control) and in softening waters.

In some waters, coagulation is poor, even with an optimum dose of coagulant. In these cases, a coagulant aid is used. The two most common coagulant aids are clay and polyelectrolytes. Clay (bentonite, fuller's earth or other clay) is useful in some waters, which are deficient in negatively charged flocs. The clay colloids provide nuclei for the formation and growth of flocs, as well as a medium to add some weight to the particles. In some cases, the adsorptive qualities of the clay may be useful in removing colour, odour and taste.

Polyelectrolytes are long chain macromolecules, having electrical charges or ionisable groups. Cationic polyelectrolytes are polymers that, when dissolved, produce positively charged ions. They are widely used, because the suspended and colloidal solids commonly found in water are generally negatively charged. Cationic polymers can be used as the primary coagulant or as an aid to conventional coagulants. There are several advantages to using this coagulant aid; the amount of coagulant can be reduced, flocs settle better, there is less sensitivity to pH and the flocculation of living organisms (bacteria and algae) is improved.

Anionic polyelectrolytes are polymers that dissolve to form negatively charged ions, and are used to remove positively charged solids. Anionic polyelectrolytes are used primarily as coagulant aids with aluminium or iron coagulants. The anionics increase flow size, improve settling and generally produce a stronger floc. They are not significantly affected by pH, alkalinity, hardness or turbidity.

Non-ionic polyelectrolytes are polymers having a balanced or neutral charge, but upon dissolving, they release both positively and negatively charged ions. Non-ionic polyelectrolytes may be used as coagulants or as coagulant aids. Although they must be added in larger doses than other types, they are less expensive.

Compared with other coagulant aids, the required dosages of polyelectrolytes are very small. The normal dosage range of cationic and anionic polymers is between 0.1 and 1 mg/l. For non-ionic polymers, the range is between 1 and 10 mg/l.

9.5 Operation of the coagulation and flocculation process

There are three fundamental steps in operating the coagulation and flocculation process:

(1) Selecting the chemicals;
(2) Applying the chemicals; and
(3) Monitoring the process efficiency.

The selection of chemical coagulants and aids is a continuing program of trial and evaluation, normally using the Jar Test (see below). When selecting chemicals, the following characteristics of the raw water to be treated should be measured:

- Temperature;
- pH;
- Alkalinity;

- Turbidity; and
- Colour.

The Jar test, although still the most widely used coagulation control test, is subjective and depends mostly on the human eye for evaluation and interpretation. The operator should measure the pH, turbidity, filterability and zeta potential to gain further information concerning the coagulation and flocculation process.

However, Jar test results help to determine the type of chemical or chemicals and their corresponding optimum dose to be used. Jar test results are expressed in mg/l. This unit must be converted to the equivalent full-scale dose in kg/d or m³/d. The treatment plant operator subsequently applies the chemical to the water by setting the manual or automatic metering of the chemical feed system to the desired dosage rate.

Although, the Jar test provides a good indication of the expected results, full-scale plant operation may not always match these results. Actual plant performance must be monitored for adequate flash mixing, gentle flocculation, adequate flocculation time, and settled and filtered water quality.

The zeta potential test is increasingly being used at water works to help determining the best pH and dosage for cationic polymers and cationic coagulants such as Al^{3+} and Fe^{3+}. The control procedure requires monitoring of the zeta potential of the coagulating water, and changing the chemical dosage when the zeta potential varies outside a range known to produce the lowest turbidity. The range, though variable from plant to plant, is between 6 and 10 mV.

9.6 Rapid mixing

The first contact of the coagulant with the water is the most critical period of time in the whole coagulation process. The coagulation reaction occurs quickly, so it is vital that the coagulant and colloidal particles come into contact immediately. After the coagulant has been added, the water should be agitated violently for several seconds to encourage the greatest number of collisions possible with suspended particles.

There are three principal types of mixer used in the rapid mixing stage: mechanical devices, pumps and conduits, and baffled chambers. Because of their positive control features, propeller, impeller and turbine type mechanical mixers are widely used. The detention times in these chambers are very short, usually <60 s (typically 30 s). Mechanical mixers can also be mounted directly in the pipeline (in-line mixers). Unlike pump and conduit type mixers, the in-line mixers can be adjusted to provide the correct degree of mixing. Since the in-line mixer requires no special tank, it is a low-cost device increasing in popularity.

9.7 Flocculation

Once the colloid has been destabilised, the growth of flocs due to the agglo-meration of the colloidal particles is promoted and enhanced by gentle mixing. This flocculation process occurs mainly in two stages known as perikinetic and orthokinetic flocculation.

In the first stage (perikinetic flocculation), particles collide and stick together as they move randomly under the influence of Brownian motion. The time taken for particles to grow so large that they are no longer significantly affected by Brownian motion depends on the frequency of collisions. The opportunity for collision is proportional to the concentration of the particles, so perikinetic flocculation is more rapid in concentrated suspensions. The time taken for effective completion of this stage is usually <60 s.

In the second stage (orthokinetic flocculation), particles are moved together by gentle motion of the water. The rate of flocculation depends on particle nature, size and concentration, and the velocity shear gradient of the water. Many mathe-matical models of this process have been developed, although they are all difficult to apply in practice. The rate of flocculation being proportional to:

- The velocity shear gradient;
- The volume of the particle zone of influence; and
- The square of the numerical concentration of particles.

Although the initial rate of floc formation is proportional to the velocity shear gradients, high velocity gradients can cause large flocs to be ripped apart as a result of internal tension, surface shear stress erosion, or both. The agitation necessary to induce flocculation is created in several ways:

- Hydraulic agitation: jets and baffled channels.
- Air injection: rising bubbles.
- Mechanical agitation: paddle and reel, and turbine and propeller.
- Solids contact: sludge blanket.

Chapter 10

Sludge blanket clarifiers

10.1 Introduction to sludge blanket clarification systems

Sludge blanket clarifiers are widely used in water and wastewater treatment, and can also be applied as a pre-treatment step located before a constructed wetland. The sludge blanket effect is obtained simply by allowing chemically pre-treated water (i.e. coagulant added) to flow upwards in a clarification unit. As water rises in the clarifier, its velocity progressively decreases until, at a given level, the upflow velocity of the water equals the hindered settling velocity of the floc suspension (e.g., the metal hydroxide precipitate). The flocs are therefore suspended in the water to form the sludge blanket.

The principal elements of the sludge blanket clarification system are as follows:

(1) Rapid mixing unit in which the raw water is mixed with coagulants, and then subjected to a delay period prior to entering the blanket;
(2) Inlet system to the clarifier;
(3) Sludge blanket and flocculation zone;
(4) Supernatant clear water zone;
(5) Excess sludge removal system; and
(6) Clarified water collection system.

10.1.1 Rapid mixing and delay time

The addition of chemicals to the raw water and their thorough mixing is the first stage in the clarification process. After the initial rapid mixing, an additional slower mixing time of a few minutes (i.e. delay time) is useful to assist perikinetic flocculation (i.e. the formation of primary flocs). This in turn results in improved orthokinetic flocculation.

10.1.2 Inlet system to the clarifier

The nature of the inlet flow to the clarifier is vital to the hydraulic stability of the blanket, the distribution of flow through the blanket and hence settled water quality. It is important to ensure verticality and centring of the inlet. Poor inlet design gives rise to short-circuiting and streaming of the water to one side of the tank. This will disturb the stability of the sludge blanket with subsequent floc carry-over in the effluent. The inlet flow must be dispersed uniformly over the inlet level. A uniform flow distribution to the blanket is vital to ensure its stability.

A zone of high turbulence at the inlet is required to maintain the blanket in a fully fluidised condition. The high energy of the incoming water is reduced in the fluidised bed by the turbulent dissipation of energy and because of the large concentration of solid particles.

A low inlet velocity would be sufficient to prevent sedimentation, whereas an inlet velocity greater than that required would provide excess energy in the incoming water. Excess energy largely increases the instability conditions of the sludge blanket. This demonstrates that even small deviations from the optimum inlet velocity have a large detrimental effect on the performance of the sludge blanket.

Four different inlet distribution systems have been used in practice, namely:

(1) Perforated plate distribution system;
(2) Point inlet distribution system;
(3) Perforated pipe distribution system; and
(4) Slot distribution system.

The inlet of the perforated plate distribution system does not produce an adequate uniform flow distribution due to clogging of the space beneath the plate. For this reason, it is not widely used.

The point inlet distribution system tends to cause excess incoming energy and the resulting blanket tends to be unstable. This excess energy could be reduced by using a multiple point inlet system, but too many inlets present difficulties in providing a uniform distribution of water.

When using the perforated pipe distribution system, the energy of the incoming water is reduced substantially and the stability of the blanket increases. However, a large number of pipes should be used to obtain the optimum inlet velocity. In such a pipe system, a uniform flow distribution is difficult to maintain due to clogging of the orifices and pipes by flocs. Pulsation can be used to accomplish a uniform distribution and prevent clogging.

In the slot distribution system, a uniform distribution of coagulated water is provided by means of a narrow slot. For large clarifiers, more than one slot inlet can be used. The dimensions of the inlet slot can easily be designed to obtain the optimum velocity. The adaptability of the size of the inlet slot is one of the greatest advantages of this type of system.

10.1.3 Sludge blanket and flocculation zone

The sludge blanket is a fluidised bed formed by the accumulation of flocs. While the top of the blanket is very often well defined, there is no distinguishable bottom of the blanket. This is because of the varying density of the floc particles forming the blanket. For floc particles to form a blanket, the upflow velocity of the water must be equal to the hindered settling velocity of the floc suspension. In a suspension, the floc particles mutually hinder one another and the fall velocity of the suspension as a whole is less than that of the individual particles (Eq. (10.1.1)).

$$v_s = v_p (1 - kc)^n \qquad (10.1.1)$$

where:

v_s = settling velocity of the suspension;
v_p = settling velocity of an individual floc;
c = concentration of the suspension; and
k and n = constants for a particular suspension.

The specific feature of a fluidised bed in a sludge blanket clarifier is that the suspended particles constituting the sludge blanket are not discrete uni-sized particles, but a floc suspension subject to aggregation and break-up. The process of flocculation involves both floc growth and break-up, so even under equilibrium conditions, there will be a range of floc sizes. The growth and break-up of the flocs depend on the growth properties of the flocs, their shear strength and on the turbulence spectra within the sludge blanket clarifier.

The mechanism of clarification within a sludge blanket tank is complex, involving flocculation, entrapment and sedimentation. Flocculation, as well as being time dependent, is also greatly influenced by the volumetric concentration of the particles and the velocity gradient in the liquid at the particle surface. With the large concentrations of particles associated with sludge blankets, the influence of time and velocity gradient are substantially reduced.

A process of physical entrapment by flocculation and agglomeration, akin to surface capture in deep bed filtration, occurs throughout the sludge blanket, but the principal mechanism in a sludge blanket is a process of mechanical entrapment and straining, which occurs when rising small particles cannot pass through the voids between the larger particles, which comprise the bulk of the sludge blanket.

The seperability of the coagulated floc suspension is governed by the physical and chemical properties of the flocs; principally by size and shape, density, shear strength, and surface conditions. For separation processes involved in sludge blanket clarifiers, the aggregational and adhesion properties of the flocs and their resistance to shear forces are of utmost importance. The aggregational and adhesion conditions of the floc suspension are determined by the local forces acting between interacting surfaces and the resistance to shear forces, the chemical nature and physical structure of the flocs.

One of the principal hydraulic requirements is that flow across any horizontal layer in the clarifier should be uniform. Non-uniform flow in the blanket causes short-circuiting within the blanket and thus instability. Some sludge blanket clarifier designs have been criticised for the tendency of the blanket to 'boil'. This can happen particularly in hot weather or where the tank is subjected to high winds.

In general, it is beneficial to have a deep sludge blanket. Greater blanket depth provides a greater dissipation of turbulence and blanket stability. Blanket stability will enhance blanket concentration, thereby improving clarification by entrapment. Increased depth will also increase the contact time of the water with the sludge blanket, enhancing chemical and physical reactions.

Solids concentration within the blanket increases with distance from the bottom, highest concentration being at the top of the blanket. The size of the flocs in the blanket also increases with distance from the inlet. Flocs at the inlet are small due to high turbulence, but since the velocity gradient reduces with movement upwards through the blanket, the flocs become larger.

10.1.4 Supernatant clear water zone

When water exits from the sludge blanket, it enters a clear water zone. This zone serves primarily as a buffer to allow surges in blanket volume to be accommodated. If flocs escape from the surface of the blanket, their free settlement velocity in the clear water zone will exceed the upward velocity, and thus they will return to the blanket. It is essential to have a uniform distribution of flow in the clear water zone. Flow conditions in this zone depend on the depth of the zone, the location of the clear water collection weirs and the location of the sludge removal devices. The clear water zone must have a minimum depth equal to half the spacing between the water troughs.

10.1.5 Excess sludge removal system

Effective removal of excess sludge is very important for the efficient operation of clarifiers. The continuing addition of coagulants to inflowing water causes an increasing volume of flocs in the sludge blanket tank. This will cause the blanket volume to increase. It follows that for steady state conditions in the clarifier, the rate of removal of the sludge must balance the rate of sludge formation. The rate of sludge formation equals the dosage rate of a coagulant, if it is assumed that the volume of the natural particles in a suspension of water is insignificant compared with that of the hydrolysis products.

Therefore, it is possible to predict the volume of sludge to be removed, on average, as a function of the coagulant dose. The removal of flocs is usually required with a minimum removal of water. Consequently, it follows that sludge removal should take place where the blanket is most concentrated. Withdrawal of sludge at any other level is inefficient, and withdrawal at the top gives positive control of the sludge blanket level.

An effective method of sludge removal consists of one or more flexible cones suspended in the water and connected to a load cell via a cable. The load cell is sufficiently sensitive that when the weight of the sludge reaches a preset value, then it initiates the opening of the desludging valve on the cone outlet. The removal of excess sludge can also be accomplished by a submerged weir, over which the sludge flows into a sludge concentrator. The weir crest level is set at the upper boundary of the sludge blanket.

10.1.6 Clarified water collection system

The clarified water is collected by a series of overflow weirs. Outlet ducts should not cause biased flow patterns. As it is the case for the inlet design, poor outlet design causes short-circuiting and streaming in the blanket, which affects its stability. The collection at the outlet must be uniform over the outlet cross-section. In normal practice, uniformity of weir loadings is accomplished by using castellated weirs. In some instances, it might be necessary to incorporate baffles in the clear water zone to minimise wind-generated surface movement passing down to the blanket and causing disruption.

10.2 Types of sludge blanket clarifier

10.2.1 Hopper-bottomed tank

The earliest type of sludge blanket clarifier was the hopper-bottomed tank. This type of tank was first developed during the 1930s and 1940s, and was used mainly in places where there was a British influence. They are usually square in plan (occasionally circular). The hopper shape in the bottom of the tank assists in the flow distribution. It has been found that square plan hopper tanks deteriorate in performance at lower upflow rates, if compared to circular tanks. However, the square plan permits an economical construction.

10.2.2 Flat-bottomed tank

There has been a continuous striving towards simplifying the shape of the tank to reduce construction costs. This is usually achieved by using a flat-bottomed tank.

In this type of design, chemically treated water is directed down to the base of the tank through suspended pipe chandeliers before passing upwards through the blanket to the surface collection system. The reversal of flow, which takes place at the base of the tank assists chemical flocculation in the same way as in the hopper-bottomed tank.

10.2.3 Pulsator

The pulsator clarifier represents the industrial application of laboratory observations and theoretical considerations. It permits the use of high upflow rates, which can be up to 8 m/h, depending on the type of suspended matter.

The clarifier consists of a flat-bottomed tank with a series of perforated pipes at its base to distribute the raw water uniformly over the entire bottom of the clarifier. A further set of perforated channels is provided at the top of the clarifier to collect the clarified water evenly and prevent any velocity irregularities in parts of the tank.

Various methods are available for admitting water via the bottom pipe system at intervals, but they all require storage of a certain volume of raw water for a certain period of time; this water then being introduced into the tank as quickly as possible.

The most economical method of effecting this operation is to introduce the raw water into a vacuum chamber from which the air is exhausted by a vacuum pump removing an air flow of about half the maximum water flow to be treated. The vacuum chamber communicates with the bottom pipe system of the clarifier.

Under these conditions, the level of raw water gradually rises in the vacuum chamber. When it reaches a value between 0.6 and 1.0 m above the level of water in the tank, a contact operates an electrical relay for quick opening of an air inlet valve. Therefore, atmospheric pressure is immediately applied to the water stored in the chamber, and the water rushes into the tank at high speed. These appliances are generally set so that the water in the vacuum enters the settling tank in 5–10 s while the filling time for this chamber is between 30 and 40 s. Air is drawn from the vacuum chamber by a fan or electric blower acting as a vacuum pump. The valve connecting the vacuum chamber to the atmosphere is opened and closed when the water reaches the upper or lower level.

The header in the bottom of the settling tank has a large cross section to reduce head loss. The orifices in all branch pipes are placed so that a uniform sludge blanket forms in the lower half of the settling tank. This blanket is subject to alternating vertical movements, and tends to increase in volume, because of impurities brought by the raw water and the flocculation reagents added. Its level gradually rises. A particular zone of the tank is used to form hoppers with inclined floors into which the excess sludge discharges and subsequently concentrates. Sludge draw-off is carried out intermittently by the discharge pipes.

The equipment has no mechanical sludge stirring system likely to break up the floc already formed. Due to the high concentration in the sludge blanket, and its buffer action, faulty adjustment of the treatment dosing rate, or a variation in the pH of the raw water, have no immediate negative effect. A slow variation in the turbidity of the clarified water is observed, but without any massive loss of the sludge in the tank.

10.2.4 Plate-type pulsators

By fitting plate modules in a pulsator above the sludge blanket, the quality of the clarified water can be improved, if compared with results of a conventional tank. The plates or tubes are usually made of plastic, and are inclined at about 60° to the horizontal. The floc particles, which escape from the sludge blanket, settle on the lower surfaces of the plate modules and accumulate there. A thin film develops until cohesion causes the film to slide back into the blanket.

10.2.5 Super pulsator

The super pulsator combines the advantages of sludge contact settling, pulsation of the sludge blanket and plate settling. It has many features in common with the pulsator, from which it has been developed, and gives improved performance. The principle of raw water feed and distribution at the base of the equipment has been retained.

The flocculated sludge rises vertically in parallel streams crossing the deep zones situated between the bottom distribution pipes and the inclined plates, which are therefore uniformly fed with water. The stilling baffles, used in the pulsator, can be omitted in the super pulsator.

The flocculated water, distributed evenly by the distribution system, then penetrates the system of parallel plates inclined at 60° to the horizontal and perpendicular to the concentrator. The underside of each plate is fitted with deflectors, which act as supports and create slow eddying movements.

The deflector plates in the sludge blanket enable a high concentration to be maintained, twice that in a pulsator operating at the same rate. The high concentration in the sludge blanket, which can reach 50% by volume, enables the super pulsator to play the role of a filter. This is a major advantage for clarifiers with deep and concentrated sludge blankets.

As with the pulsator, the top surface area of the sludge blanket is limited by the provision made for decanting sludge into the concentrator zone where no force is exerted by the upward movement of water. The recovery of clarified water is achieved by a system of collectors. The flexibility of operation of the super pulsator permits very quick starting.

10.3 Plate settling in sludge blanket clarifiers

In a sludge blanket clarifier, the use of plate modules improves the quality of clarified water at the same upward velocity by trapping the residual flocs, which escape from the blanket. Conversely, for the same quality of clarified water, the volume of water treated by the equipment can be considerably increased.

If inclined plates are introduced into a sludge blanket maintained in suspension, an accelerated settling process is observed. The sludge settles onto the lower plate, and is then subjected to a downward flow towards the base of the sludge blanket. At the same time, the water released by this process gathers under the upper plate and tends to rise rapidly towards the top of the tank. Very high settling velocities can be achieved, which can reach 20 m/h, depending on the nature of the water and the type of floc.

Numerous types of plate type static settling tanks, which are equipped with parallel plates or tubes, exist. Some systems combine a mixing unit, accelerated flocculator fitted with a super pulsator, deflector plates and a plate-settling zone. The sludge falls by gravity, and is concentrated in a hopper at the bottom of the tank.

Chapter 11

Flotation systems

11.1 Flotation using blown air

Flotation can be spontaneous when the specific mass of particles to be eliminated is less than that of water, or it can be induced by the artificial fixation of bubbles of air or gas onto the particles to be removed, giving them an average specific mass less than water. Flotation using blown air is effectively spontaneous flotation improved by blowing bubbles of air (several millimetres in diameter) into the liquid mass. For this purpose, porous devices are used to form the bubbles.

If necessary, where heavily concentrated liquids are involved, medium size bubble diffusers are arranged, so as to produce local turbulence designed to encourage the division of the air bubbles. The size of the bubbles must be small enough to enable them to adhere to the particles to be floated.

Two separate zones are generally provided in plants designed to remove light materials (e.g., grease, thick oil, large fibres and paper); one zone for mixing and emulsifying, and the other, a calmer zone, for flotation. In the emulsion zone, the suspended solids are stirred and inter-mixed with air. The path of the bubbles is increased by the spiral flow thus created. In the separation and collection zone for floating materials, the cross flow is very slow and turbulence is reduced.

11.2 Flotation using dissolved air

It is standard practice to reserve the term flotation for processes using very fine air or microbubbles between 40 and 70 μm in diameter, similar to those present in white-water running from a tap on a high pressure main. Separation by flotation of solid particles in a liquid suspension obeys the same laws as sedimentation, but in a reverse field of force.

There is a variation in the upward flow rate of air bubbles according to their diameter. Twenty micrometre bubbles have an upward flow rate of several millimetres per second, whereas bubbles of several millimetres in diameter have between 10 and 30 times greater velocities. If, for a given volume, an emulsion is introduced at one point with an outlet provided at the other end, the period of immersion of the air bubbles in the water, and similarly the space filled with bubbles, will be greater, the smaller the upflow rate of the bubbles and their corresponding diameter.

For a flotation unit of a given cross section, the use of bubbles several millimetres in diameter will result in an air flow much greater than for microbubbles, if a satisfactory distribution of the bubbles is desired over the cross section. At the same time, this increase in air flow will set up turbulent currents, disturb satisfactory operation, and induce mechanical mixing.

The bubbles have a flotation effect only to the extent to which they adhere to the particles. This generally assumes that their diameters are less than the diameters of the material or floc in the suspension.

A flotation process using means other than microbubbles can only be used in suspensions containing light and bulky materials whose surface deposits are not disturbed by whirling movements. The application of flotation in the field of wastewater treatment is as follows:

- Separation of flocculated matter in the clarification of surface water (in lieu of settling, generally for which the water is cold, with a low mineral content and charged with organic matter);
- Separation of flocculated or non-flocculated oil in wastewater from refineries, airports and steelworks;
- Separation of metallic hydroxides or pigments in the treatment of industrial wastewater; and
- Thickening of activated sludge (or mixed activated sludge and primary sludge) from organic wastewater treatment plants before application of constructed wetlands functioning as sludge drying beds, for example.

The techniques vary according to:

- The method used for bubble formation;
- The method of feeding the flotation unit;
- The shape of the structures; and
- The method of collecting the skimmings.

The most widely used technique for producing microbubbles is pressurisation. The bubbles are obtained by the expansion of a solution enriched with dissolved air at several bars of pressure. The pressurised liquid used is either raw water or re-circulated treated water. The rate of flow of the pressurised water is generally only a fraction (10–30% of that to be treated) of the rate of flow of the plant, at pressures between 3 and 8 bar. On average, about 60% of the excess air is allowed to dissolve in relation to the rate of saturation at atmospheric pressure. Hence, the compressed air consumption varies between 15 and 60 l/m^3 of water to be treated.

When the quantity of material to be floated is high, and thickening is particularly desired, as in the case of activated sludge, the recycled flow can reach 200% of the nominal flow of the flotation unit. It is possible, with the use of polyelectrolytes, to obtain sludge concentrations between 3 and 6% for specific loadings of 5 to 13 kg of dry solids/m^2/h at a downward velocity of 2 m/h.

Electroflotation is another technique, in which the objective is to produce bubbles of hydrogen and oxygen by electrolysis of the water by means of appropriate electrodes. The anodes are highly sensitive to corrosion and the cathodes to scaling by carbonate removal. When the protection of anodes is required, protected titanium electrodes may be used, or the periodic reversal of the electrodes may be employed to effect self-cleaning. A preliminary treatment of the water or periodic descaling of the cathodes must be allowed for.

Flotation is often combined with preliminary flocculation. By incorporating a flocculation aid, the floc can be enlarged and the particle surface area increased. This gives improved adhesion of the bubbles and an increase in the upward flow rate of the floc.

The separation or downward flow rate of the water used in flotation units varies according to the nature of the suspensions to be treated, and also according to the method of generation and distribution of the microbubbles. For a given flotation unit, the downward velocity and concentration of the sludge floated are strongly influenced by the value of the ratio of 'quantity of air dissolved' over 'quantity of material to be floated'.

The greater this ratio, the greater the upward force given to the particles; the higher the downward velocity, the lower the sludge density, and the greater its concentration in dry solids.

This ratio cannot exceed the upward flow of the bubbles. Although, as stated above, an extremely fine size of bubbles is particularly conducive to their distribution over the entire surface, and thus separation efficiency. However, this can sometimes limit the traverse speed, and therefore limit the output of water treated by the plant.

11.3 Flotation units

11.3.1 Technology

Flotation units can be circular or rectangular. Rectangular units are usually employed for the treatment of water and wastewater, as they can be constructed as a monobloc together with the flocculator and filters, so that land requirements are minimal.

With a flotation problem to solve, preliminary tests in the laboratory or a test on a semi-industrial scale are recommended. It is then advisable to proceed with the design calculation stage, which is governed by two main variables: the downward velocity and the quantity of solids to be floated per unit area and time.

A low downward velocity, while reducing the particle flow to the bottom of the tank and theoretically leading to a higher separation efficiency, also leads to an increase in the retention time of the floating sludge. This may cause deterioration and splitting-up of flocs, resulting in the reduction of the sludge concentration. It falls to the specialist to find the best compromise between sludge thickening and separation efficiency. Furthermore, the internal layout has considerable influence on the performance of the flotation unit.

For distribution of air bubbles, circular shaped flotation units are superior to rectangular units. The distance between the inlet column and the outlet funnel is shorter for the same capacity, and an almost uniform distribution of the bubbles can be maintained over the entire horizontal section of the unit.

11.3.2 Water feed

The water feed always incorporates a column or chamber with a dual function:

- It brings into mutual contact the water (which may or may not be flocculated) to be treated and the pressurised water. The water should be depressurised immediately upstream of the unit (i.e. bubbles should not be wasted); and
- It disperses the kinetic energy of the raw water and the pressurised water mixture, and reduces the speed before the process water reaches the actual flotation zone.

The chamber also enables any large bubbles that may have formed upstream to be removed immediately. The points and respective levels of introduction of raw and pressurised water are of major importance in the case of prior flocculation.

Precautions to be taken in calming the mixture, when it is introduced into the flotation zone, vary according to the size, stability and density of the flocculated or non-flocculated substances to be floated. In the case of the flotation of slightly viscous oils, this point should be particularly watched carefully.

The emulsified water is generally introduced in the top half of the unit. Sludge is collected at the open surface. The clear liquor outlet is fitted in the lower third of the unit. The deeper the unit and the greater the quantity of sludge that can be deposited, the further away will be the clear liquor pick-up from the bottom. Collection is usually carried out in a peripheral chamber. In certain cases, immersed headers and laterals can be used. The more uniform the distribution of water and microbubbles, the shallower (generally between 2 and 4 m) the flotation unit can be designed.

11.3.3 Formation of bubbles

Several devices require the introduction of air upstream of the pressurisation pump. Systems incorporating an air cushion saturation tank are more costly, but their operation is very stable. Pressurised air dispersal devices in the pump discharge pipe have also been proposed. For electroflotation, a bank of electrodes covers the surface area of the tank.

11.3.4 Collection and removal of sludge

The surface sludge layer can in certain cases attain a thickness of many centimetres, and be extremely stable (e.g., thickening of activated sludge). In other cases, it is thinner and more fragile (e.g., flotation of flocs of metallic hydroxides or of oils). When the sludge removal is not total, the layer thickens and, with time, acquires a degree of cohesion that facilitates the adhesion of floating particles.

The gradual and regular removal of sludge is an important point. The scraper mechanism must be especially strong, if it is fitted in a floatation unit designed for sludge thickening. On circular flotation units, one or more scrapers push the sludge into a radial collection channel with a length equal to half the radius. The access ramp must be constructed in such a way that contact with the scraper is always ensured.

The number of scrapers is governed by the quantity of sludge to be removed, by the rapidity with which this removal must be carried out to avoid the risk of deaeration, and by the distance, the sludge can be pushed without breaking up. In rectangular flotation units, the sludge is pushed by a series of scrapers driven by endless chains to a removal channel situated at one end.

Chapter 12

Slow filtration

12.1 Introduction

The processes within most filtration systems are similar to those in wetland systems. Moreover, traditional slow sand filters are operated similar to vertical-flow constructed treatment wetlands.

Filtration is the process by which water is separated from suspended and colloidal impurities. The number of bacteria is reduced, and changes in the chemical characteristics of the water are brought about by passing it through a porous substance or filter media. In principle, this porous substance may be any material including sand, crushed stone, anthracite, glass, cinders, consolidated layer of porous concrete, stoneware and plastic.

However, sand is used almost exclusively as the filtering material, because of its easy availability, relatively low cost and good performance. On occasions, anthracite has a useful role to play in sand filters. Filtration should not be confused with straining; filtration is not simply a straining process!

12.2 Slow sand filtration

12.2.1 Elements of a slow sand filter

A slow sand filter is essentially a bed of sand through which the pre-treated water passes (downwards), together with the necessary structures to control the flow, and to remove the water after filtration. The filtration rate varies between 0.1 and 0.4 $m^3/m^2/h$. The filter bed consists of fine grains (effective diameter between 0.15 and 0.35 mm) with a total depth of about 1 m initially. As filtration progresses, the suspended and colloidal matter from the raw water are retained in the very top of the filter bed. The clogging material here may be removed and the filter restored to its original capacity by scraping off this top layer of dirty sand to a depth varying from one to a few centimetres.

The traditional slow sand filter is an open basin (2.5–4 m deep), usually rect-angular in shape, built below finished ground level and varying in area from a few hundred to a few thousand square metres, and sometimes even larger. The box is normally constructed in reinforced or pre-stressed concrete (tradition-ally masonry or earth on puddled clay cores, with sloping sides). This box is filled with a 0.6–1.2 m thick layer of sand on top of which the water to be treated is present to a depth of 1–1.5 m. At the lower end, this sand bed is supported by a drainage system, so-called filter bottom or underdrain, which at the same time allows the passage of the filtered water.

A variety of forms is available for the underdrain, the commonest being porous concrete filter underdrains. In cold climates, slow sand filters must be covered to prevent freezing in winter. In moderate and particularly in hot climates, covering is sometimes practised to prevent algal growth in the supernatant water and sub-sequent rapid clogging of the filter bed, and also to avoid bacteriological and other pollution brought into the filter by wind and birds, for example. The slow sand filter is not complete without a number of influent and effluent lines, provided with valves and controllers to keep the raw water level, and hence the filtration rate constant.

During operation, the raw water to be treated enters the filter through a valve, flows down the sand bed, through the underdrainage system, and out through a venturimeter (or other measuring device) and the regulating valve into a weir chamber. After overflowing the weir, the water is discharged through a further valve (all other valves being closed) towards the clear water well.

With the regulating valve, any rate of filtration, from the minimum to the maximum allowable value, may be obtained, while the weir on one hand prevents the filtered water level to drop below the top of the sand bed and on the other hand makes the filtration rate independent from uncontrollable variations of the water level in the clear well. At the same time, this weir provides an appreciable amount of aeration (i.e. decreasing carbon dioxide and increasing oxygen concentrations).

Good ventilation must be provided to keep the partial oxygen and carbon dioxide pressures in this chamber as near as possible to those in the outside air. This ventilation is also required for the filter, allowing air and other gases liberated in the underdrainage system to escape without hindering the movement of water.

During the filtration process, a gradual clogging of the pores in the upper part of the filter bed occurs, increasing the resistance to flow. This increase is compensated by opening the regulating valve, keeping the overall loss of head the same, and hence the filtration rate constant. When, after an extended period of service, the effluent valve is fully opened, a further increase in filter resistance would result in a lowering of the filtration rate, and the filter must be taken out of service for cleaning.

If the composition of the raw water is such that a large amount of floating material accumulates at the water surface, this scum must be removed regularly through scum outlets at the four corners of the filters, after the water level has been raised. This scum removal is particularly important near the end of the filter run, so as not to render filter cleaning more difficult than is strictly necessary.

Cleaning of a slow sand filter is commonly executed by scraping off the upper 1 or 2 cm of sand. However, this requires a dry filter bed with the water level about 0.2 m below the top. In order to drain the filter, the raw water inlet valve is closed, allowing the filter to discharge (say overnight) through further valves to the clear well. The shallow layer of water remaining on the filter bed is removed through another valve connected to a box of which one wall is built up of stoplogs (usually made of concrete), so that the top of this discharge wall can be kept near the top of the sand bed. The pore water in the upper 0.2 m of the filter bed is finally taken out.

After the cleaning operation has ended, the filter is slowly recharged with filtered water from below to a level of about 0.1 m above the top of the sand bed, allowing air accumulated in the pores of this bed to escape. The remaining depth is filled with raw water from above, taking care not to disturb the surface of the sand bed, because it is protected by a shallow layer of water only. When the raw water level in the filter has regained its normal value, the filter operates at about one quarter of its normal rate.

In the following 12–24 h, the filtration rate is slowly increased to the required value, after which a sample is taken. If this sample satisfies the quality criteria, the filter is brought back to normal operation. When the water quality is not satisfactory, the filtered water is carried to waste until the water quality is acceptable. After normal cleaning, this ripening process lasts 1 or 2 d; after repairs such as re-sanding, it may, however, take longer.

12.2.2 Mechanisms in a slow sand filter

The overall removal of impurities associated with the process of filtration is brought about by a combination of several different processes, the most important of which are sedimentation, mechanical straining, adsorption, and chemical and biological activity.

Mechanical straining is the purifying process most easy to grasp, removing the particles of suspended matter that are too large to pass through the gaps between the sand grains. As such it takes place almost entirely at the surface of the filter, where the water enters the pores of the sand bed, and is independent of the filtration rate. Even with a grain size of 0.15 mm, the smallest pores are still a little over 20 μm in diameter, and are thus unable to retain either colloidal matter (0.001–1 μm) or bacteria ($\leq$15 μm).

By the twisting movement of the water through the pores of the filter, however, velocity gradients are set up, which cause some flocculation. Some aggregation of the finely divided particulate matter will now occur and part of the flocs thus created are retained at a greater depth in the sand bed. Moreover, a slow sand filter is not immediately taken into service, but a 'ripening' period precedes the filter run. This is similar to a constructed treatment wetland.

In this period, coarser suspended solids (SS) caught at or near the surface of the bed form a very fine porous layer, the so-called filterskin or schmutzdecke (German for layer of dirt), which greatly promotes the straining efficiency and thus reduces effluent turbidity. Straining efficiency is also enhanced with time by deposits on the grains in the filter bed, which constricts the pore openings.

Sedimentation removes particulate suspended matter of finer sizes than the pore openings by precipitation onto the sides of the sand grains. For a normal filter bed (media porosity of approximately 38% and grain size of 0.25 mm), this gives a gross area for sedimentation (i.e. sand grain surface area) of approximately 15 000 m^2/m^3 of filtering material.

Even when only a fraction of this area is effective (facing upwards, not in contact with other grains and not exposed to high scour velocities), the area for deposition can easily exceed 1000 m^2/m^3 filter volume. The surface loading, as a quotient of the amount of water to be treated, and the area of deposition will now be extremely small, with a filtration rate of 0.2 m/h, and $\leq 2 \times 10^{-4}$ m/h.

Sedimentation efficiency is a function of the ratio between surface loading and the settling velocity of the particles. It follows that the settling velocity is larger than the overflow rate, and virtually complete removal by sedimentation may be expected. Small and light particles are only partly removed. Although flocculation will slightly increase the removal efficiency as the water progresses through the bed. However, truly colloidal matter will not be removed by this mechanism.

Adsorption has many facets, the simplest of which is the collusion of suspended particles and sand grains, after the former have left a streamline (flowline) by centrifugal force. The impurities will adhere to the 'sticky' gelatinous coating formed on the grains by previously deposited bacteria and colloidal matter. Much more important is the active promotion of this adsorption by the physical attraction between two particles of matter (van der Waals forces), and especially by the electrostatic attraction between opposite electrical charges (Coulomb forces).

Mass attraction is present everywhere and always, but its magnitude decreases with the 6th power of the distance between particles. For the previous example, 1 m^3 of filter sand has a gross area of 1500 m^2. With a porosity of 38%, the sand contains 0.38 m^3 of water, which corresponds to a thickness of a waterfilm around the grains of 25 μm, if spread over the surface area. This is too large to carry colloidal and molecular dissolved impurities to the sides of the grains. Purification in this way is only possible, if other actions (e.g., centrifugal force) have brought the particles in the immediate vicinity of the surface of adsorption.

Electrostatic forces are inversely proportional to the second power of the distance, and their influence consequently reaches deeper into the body of the passing liquid. By the nature of its crystalline structure, clean quartz sand has a negative charge, and thus attracts positives in the form of colloidal matter (e.g., flocs of carbonates, iron hydroxide and aluminum hydroxide) and cations (e.g., iron, manganese and aluminium). Colloidal matter of organic origin (e.g., bacteria) mostly carries a negative charge. The organic colloids are consequently not attracted, and indeed when a filter with clean sand is first taken into service, such impurities are not removed.

A slow sand filter needs a ripening period, during which positively charged particles are adsorbed. Positive charges may accumulate on some of the filter grains, so that oversaturation occurs, by which the charge on these coated particles reverses and becomes positive. This is similar to the ripening process within a constructed wetland.

After the primary adsorption during the breaking-in period, secondary adsorption is able to remove negatively charged particles, colloidal matter of animal or organic origin as well as truly dissolved impurities (e.g., anions including NO_3^- and PO_4^{2-}). When secondary adsorption leads to oversaturation, the overall charge again becomes negative, allowing adsorption of positively charged matter (and so on). Moreover, ions on the sand grains may be dragged away by the flowing liquid, by which again a reversal of charge will occur.

By all the actions mentioned above, impurities are transferred from the water to the surface of the filter grains. After some time, so much material will have accumulated that in a unit period of time, the rate of deposition will equal the rate removed by scouring. Indeed, when filters are cleaned regularly by scraping (removal of the upper 1 or 2 cm), the main body of the filter is left untouched. If no persistent fouling occurs, the filter bed will remain in place virtually for decades. This is due to chemical and microbiological oxidation, which destroys the impurities collected on the surface of the grains.

During the break-in period, bacteria from the raw water are adsorbed on the sand grains, where they multiply selectively, using as food the organic matter deposited on the grains. This food is partly oxidised to provide the energy, which the bacteria need for metabolism (dissimilation), and which is partly converted into cell material for growth (assimilation), thus converting dead organic matter into living organic matter.

The dissimilation products are carried downwards by the water to be used at greater depths by other bacteria. In this way, the degradable organic matter present in the raw water is gradually broken down into harmless inorganic salts such as H_2O, CO_2, NO_3^- and PO_4^{2-} (mineralisation) to be discharged in the finished water. With the limited amount of organic matter supplied by the incoming raw water, only a restricted bacterial population can be maintained, and the growth (assimilation) is therefore accompanied by an equivalent die-away. This again liberates organic matter, which is broken down as previously described. Thus, all degradable organic matter in the raw water is converted into mineral constituents.

The bacterial activity described is most pronounced in the upper part of the filter bed and gradually decreases with depth, as food becomes scarce. With each scraping (cleaning), a portion of the bacteria is removed, and again a ripening period is necessary to bring the population up to the required strength. Below a depth of 30–40 cm (deeper if the filtration rate is higher), bacterial activity is very low. However, chemical reactions occur at greater depths, converting microbiological degradation products such as amino acids into ammonia, nitrites and nitrates (nitrification).

In order to keep the water quality acceptable, a sufficient oxygen content of the filtered water is required, usually >4 mg/l. This is to prevent local anaerobic conditions. Under anaero- bic conditions, the products of the breakdown of organic matter are likely to include hydrogen sulphide, ammonia, manganese, iron (reduced), and various taste and odour producing compounds.

The other important variable is temperature. If the temperature drops <2°C for any long period, difficulties such as freezing may arise. The filter must be covered to prevent heat loss.

Bacteria move by themselves to the filter grain surfaces, where their food is concentrated. For intestinal bacteria, the water environment is a decidedly unhealthy place. The temperature is too low, and insufficient organic matter of animal origin is available to suit living requirements.

In the upper part of the filter bed, predatory organisms (e.g., protozoa and lower metazoa) are feeding on bacteria, while with increasing depth below the filter bed, surface food becomes so scarce as to starve bacteria, which venture this far. Although there is little data to prove it, it is practically certain that the micro-biological life in a slow sand filter produces various antagonistic actions, such as killing or at least weakening intestinal bacteria with chemical (antibiotic) or biological poisons. The overall effect is a marked decrease in *Escherichia coli*, for example, and as pathogens are still less likely to survive in adverse environmental conditions, an even larger drop in their numbers.

12.3 Algal actions

A description of the slow sand filtration process would not be complete without mentioning the role of algae, present in the supernatant water. A major part of the changes in water quality brought about by slow sand filtration is produced by these algae, which are first carried along by the raw water, and then multiply actively during the often long intervals between successive cleanings. As an autotrophic organism, algae need light for their photosynthetic processes, and hence algae are only present in uncovered filters.

The most important feature of algae is their ability to build cell material from simple minerals, with the aid of solar energy. The carbon cycle part of this process is described schematically in Eq. (12.3.1).

$$n(CO_2) + n(H_2O) + energy \rightarrow (CH_2O)_n + n(O_2) \text{ (active growth period)}$$
$$(12.3.1)$$

For their metabolism, algae need energy, which they produce by oxidising organic matter. The same reaction occurs when algae die, their cell material is liberated and consumed by bacteria in the filter bed (Eq. (12.3.2)).

$$(CH_2O)_n + n(O_2) \rightarrow n(CO_2) + n(H_2O) + energy \qquad (12.3.2)$$

Three phases, which affect the relative magnitude of algal numbers, can be identified; growing, constant and decreasing population of algae. During the active growth period, oxygen increases and carbon dioxide decreases. Bicarbonates are broken down into carbonates and carbon dioxide (Eq. (12.3.3)).

$$Ca(HCO_3)_2 \rightarrow CaCO_3 + CO_2 + H_2O \qquad (12.3.3)$$

The lowering of the bicarbonate content decreases the temporary hardness (alkalinity); the insoluble carbonates will then precipitate, promoting filter clogging. As the growth of algae continues, their volume increases, and subsequently hinders the water movement, and they need to be removed by cleaning.

Steady state conditions imply that energy is required. Therefore, the phenomenon is restricted to daylight hours (e.g., oxygen increases during the day and decreases at night). Hence, a diurnal variation in oxygen content and pH is likely to occur. This may affect waters with high iron and manganese contents, resulting in aesthetic problems with the finished water.

In temperate climates, temperature drops in autumn cause the algae to die. This liberates organic matter and subsequent biochemical degradation consumes oxygen and produces carbon dioxide. A higher carbon dioxide content will lower the pH, making water more aggressive (Scholz, 2003). So much organic matter is liberated at this time that some organics may clog the filter and necessitate cleaning.

The final beneficial factor of algal growth requiring consideration is the growth of filamentous species. These form on the surface of the filter a gelatinous mat, removing suspended matter and bacteria by straining and adsorption. Especially, the true water bacteria will multiply in this mat, creating a zooglea or schmutzdecke (see above) of bacterial slime, an adhesive medium for plankton and diatoms. Straining and adsorption efficiencies of the schmutzdecke are thus enhanced and less suspended matter will reach the filter, thus prolonging filter runs.

12.4 Summary of slow sand filtration

The advantages and disadvantages of traditional slow sand filters are similar to those of constructed treatment wetlands. The advantages of slow sand filtration are summarised below:

- No pre-treatment, perhaps other than preliminary sedimentation, is needed;
- No chemicals are needed;
- Less skilled supervision is required than for rapid sand filtration;
- It has a good bacterial removal efficiency;
- The effluent is less corrosive than that from rapid sand filtration; and
- There is a long period of satisfactory operation between cleanings.

The disadvantages are as follows:

- A large area of land is required and the structures are expensive (capital cost is high);
- It has poor colour removal efficiency; and
- Turbidity removal is poor, if the raw water turbidity is >40 NTU.

Conditions that could favour the installation of a slow sand filter:

- The raw water is moderately polluted, but has a low turbidity;
- A location where chemical supplies are unreliable;
- A shortage of skilled labour to operate rapid sand filters (Chapter 13); and
- Abundant labour force for construction and cleaning.

Recent outbreaks of waterborne giardiasis (*Giardia lamblia*) have demonstrated the need for an effective technique for removal of *Giardia* cysts (*Giardia* is a protozoa). These cysts can persist in low temperature and low turbidity waters for many months, and are very resistant to the action of disinfectants. Slow sand filtration, after ripening, has been shown to be highly effective at achieving complete removal of *Giardia* cysts (Chapter 21).

Furthermore, *Cryptosporidium parvum* is another organism causing acute diarrhoea illness in man. It forms oocysts, which are not affected by chlorine, but are, however, removed by slow sand filtration (Chapter 21).

Chapter 13

Rapid filtration

13.1 Elements of a rapid sand filter

A rapid sand filter (also called a rapid gravity sand filter) can be located before a constructed treatment wetland. The mechanisms of filtration in a rapid sand filter are essentially the same as those in a slow sand filter, except that the biological processes are minimised. This results from the much shorter filter run times between cleanings, preventing the establishment of mature biological growths.

When rapid sand filters were initially introduced, the filtration rates (loading rates) were between 3 and 4 $m^3/m^2/h$. At these relatively high rates of filtration, the following observations were made:

- Coagulation was required in most cases to prevent the impurities from being drawn deep into the filter bed;
- Surface cleaning was no longer adequate, because impurities were drawn deep into the sand bed; and
- Clogging occurred much more rapidly with filters requiring cleaning at intervals between 2 and 3 d instead of intervals between 30 and 100 d for slow sand filters.

For rapid sand filtration to be acceptable, it was necessary to develop a method of cleaning the full depth of the filter, rapidly and economically. The method adopted was to remove the impurities from the sand bed by a reverse flow of water, either preceded or accompanied by some form of agitation to loosen the impurities from the sand grains. Moreover, a rapid sand filter comprises the following construction elements:

- A sand bed in which filtration occurs;
- A support for the sand bed;
- An underdrain system to remove filtered water, and to admit backwash water (and air for agitation, if used);
- An inlet for water;

- An outlet for used wash water; and
- Means for controlling the flow through the filter.

In normal operation, the inlet valve and filtered water valve are open, and all other valves are closed. Water enters through the inlet valve, passes down through the sand and subsequently the underdrain system, and out through the filtered water valve.

To start a backwash cycle, using air scour followed by water scour, the inlet valve is closed, but filtering is allowed to continue for a time to avoid excessive loss of settled water. After a reasonable drawdown period during which some of the settled water in the filter chamber is filtered instead of wasted, the filtered water valve is closed and the waste valve opened, thus dumping into the drain that portion of the settled water, which was above the level of the backwash collecting trough.

The air valve is then opened to admit air under pressure into the underdrain system where it is distributed evenly underneath the filter. The air bubbling upwards through the sand bed causes agitation, which loosens the impurities from the sand grains. At the end of the air scour, the air valve is closed and the backwash valve opened to admit water to the underdrain system at a rate sufficient to wash the sand. The water passes upwards through the sand and carries with it impurities into the backwash collection troughs, and hence into the drain.

In order to finish the backwash cycle, the backwash and waste valves are closed, and the inlet valve is opened. When the water has reached a satisfactory level in the filter basin, the filtered water valve is again opened and filtering is resumed.

13.2 Sand bed

The sand bed in a conventional rapid sand filter consists of clean silica sand (depth between 0.60 and 0.75 m). The effective size of the sand used is between 0.9 and 1.0 mm. This size is required, because it is necessary to use a coagulant aid, and water is cold in winter. This arrangement produces a tough floc.

Anthracite is often used as a filter media in the USA. The crushed anthracite has a density of about 1400 kg/m^3 (compared with silica sand, which has a density of 2650 kg/m^3), and it does therefore not require high backwash rates to achieve fluidisation of the filter bed.

When a sand bed is washed, there is a tendency for stratification to occur, with the larger particles migrating to the bottom of the bed, and the smaller particles migrating to the top. In normal operation, the water is therefore filtered first through the fine sand and later through the coarse sand. The length of a filter run depends on the amount of storage voids available in the sand bed for retention of impurities. If the storage space in the fine sand layer becomes clogged before the storage space in the coarse sand layer (located deeper in the filter), the length of run is likely to be shorter than ideal.

Dual media filters aim to overcome this problem. If anthracite with a diameter of 2 mm is placed over sand with a diameter of 0.6 mm, for example, the anthracite, being lighter, will form a surface layer after backwashing, so that the water will be filtered through the coarse anthracite and subsequently through the finer sand, and ideal conditions are nearly approached.

Mixed media filters use three different materials. The bottom layer of the filter consists of fine garnet (density $= 4200$ kg/m^3), the middle layer of silica sand (2650 kg/m^3) and the upper layer of coarse anthracite (1400–1700 kg/m^3). The three materials are not so closely graded that they form separate layers, so mixing takes place, giving a gradual gradation of void sizes from large voids near the surface to fine voids near the bottom. The sand support bed requires a layer of coarse garnet sand to prevent the fine garnet sand from penetrating the gravel bed and distributors.

The sand support commonly consists of layers of gravel placed over the filter compartment floor. The purpose of this support is threefold:

• To provide uniform drainage conditions for removing filtered water from the sand bed;
• To prevent sand from entering the underdrainage system; and
• To assist in the even distribution of backwash water.

In earlier filters, the bottom layer of gravel consisted of material with a diameter between 50 and 70 mm, but the trend nowadays is towards smaller gravel in the bottom layers; a maximum of between 15 and 25 mm is often used. The total thickness of gravel is between 300 and 450 mm for many installations. The uppermost layer of the gravel must prevent the intrusion of the sand; it can consist of coarse sand with an effective diameter between 1 and 2 mm. With mixed media filters, the uppermost layers of the support consist of (high-density) coarse sand. Much attention should be given to the design of the gravel layers to ensure even backwash flow, effective cleaning of the sand and freedom from gravel displacement.

13.3 Underdrain system

The underdrain system is hydraulically designed to carry the backwash water. The backwash water flow rates through sand beds in different systems range from about 6 to 16 l/s/m^2, whereas filtering rates are only between 1.3 and 4.1 l/s/m^2. The backwash water and the air for scouring the sand are distributed as evenly as practically possible over the full extent of the sand bed.

Where the distribution nozzles are used as controls, the flow distribution will be uneven unless the combined head loss through the nozzles, gravel bed and fluidised sand to a depth equal to the original depth of the sand bed is a monotonic increasing function of the upflow rate.

The total head loss in the distribution channel should be ≤10% of the nozzle head loss, so that the backwash flow rates through the nozzles should not vary by ≥5% as a result of friction losses in the underdrains. Too much coarse gravel in a gravel bed will permit free horizontal movement of backwash water in the space between the nozzles and the sand bed, and therefore partially negate the distribution value of the nozzles.

Where air scour is used in conjunction with a drainage system, which uses nozzles, the nozzles are specially designed to distribute the air evenly. These nozzles are fitted with hollow stems, which project down into the laterals or into the plenum. A small air control orifice is drilled in the stem of each nozzle just below the filter floor. The pressure difference, which can be sustained across this orifice, is dictated by the length of the stem below the hole. The greater the allowable pressure difference, the less is the effect of extraneous disturbing influences on the air flow rate. A hood over the top of the nozzle can help to prevent water from entering the top of the stem and interfering with the air distribution. During the air scour cycle, air is introduced into the underdrain system; it escapes through the nozzles into the filter.

If water and air are to be used simultaneously for filter cleaning, care must be taken to ensure that the necessary water can be distributed to the various parts of the filter in such a way that the height of the waves in the water surface is small in comparison to the head used for forcing the air through the control orifice (otherwise, uneven air scour will result). This condition can best be attained in a plenum distribution system, particular care being taken to dissipate the energy of the incoming water, which could cause excessive waves and turbulence.

For the low head loss arrangement, backwash water can freely enter the lower gravel layers. The head loss, which controls the flow distribution, occurs in the upper gravel layers. If air scour is required, air is supplied through a separate layer of perforated air pipes set in one of the gravel layers.

The washwater is forced up through the filter bed and emerges laden with impurities removed from the sand. The troughs to collect the dirty water are placed as low as practically possible, in view of the need to avoid excessive disturbance of the hydraulic flow pattern in the expanded sand bed. The troughs discharge the water into the gullet, which is the main collection channel.

If it is expected that some of the displaced impurities may be too heavy or may settle too rapidly to be carried up into washwater collection troughs. The gullet wall may be used as a weir. The depth of water over the sand during the backwash is kept to a minimum, so that the horizontal velocity thus induced may carry the impurities over into the gullet. This is known as a cross-wash system. In order to intensify the horizontal velocity, water is sometimes introduced at the surface of the sand on the side opposite the gullet. Backwash procedures are equally as important as the design of the underdrain, sand support and sand bed in planning a system, in which the sand is thoroughly and evenly washed, without the occurrence of mud balls and gravel movement.

A filter backwash is started in response to one or more of the following conditions: excessive head loss, effluent turbidity or run time. In most installations, the head loss criterion is the most frequent reason for backwash initiation; the usual head loss limit is set at about 2 m. If the floc in the water is too weak, it sometimes happens that it is not retained in the filter bed until the desired maximum head is attained, and a turbidity breakthrough occurs.

In conditions where the pretreatment by clarification is very good, or the follow-through rates are low, the floc trapped in the sand bed may become so well gelled with time that thorough washing would be difficult, if the operator was to wait until the usual maximum head loss was reached. In such a case, the time limitation would control the initiation of backwashing.

Common practice has been to use an air scour method in which air is bubbled through the sand bed at a rate between 5 and 15 $l/s/m^2$ for a period of about 300 s (to let friction dislodge the impurities from the sand grains), followed by a low to medium-rate water backwash at 7–10 $l/s/m^2$ until the water becomes clear after about 600 s.

A recent trend is to use a period of combined water and air scour during which the air loosens the impurities, while a low-rate water backwash of about 6 $l/s/m^2$ carries them to the surface and prevents their penetration, which sometimes occurs with air scour only. This period is then followed by a medium-rate backwash at a rate between 9 and 12 $l/s/m^2$ to clear out the loosened impurities. Satisfactory combined water and air scour can be achieved, only if particular attention has been paid to the design of the water and air distribution systems.

In hot climates, where the raw water may be polluted with nutrients, the growth of algae in the sedimentation and filter basins may be a problem. This can be overcome by chlorination of the water before sedimentation. The viscosity of the warmer water found in hot climates is lower than that in temperate climates, and therefore the backwash rate needed for efficient filter cleaning is greater than that with cooler water.

The flow control for rapid sand filters is achieved in three different ways. Most systems include some means of automatic flow control; control valves operated by signals from level-sensing or flow-sensing elements. Care should be taken to avoid control conditions, which lead to controller instability, such as 'hunting' caused by continual over-correction.

Flow control systems in water treatment plants are usually operated hydraulically or pneumatically. The damp conditions often occurring in filter control galleries are not conducive to the reliable long-term operation of electrical equipment.

In a treatment plant equipped with downstream flow control, a flow measuring device (e.g., venturi, orifice or weir) is used to provide a signal, which operates the flow control valve. The control system adjustment may be set by hand, so that the rate of flow is kept close to a pre-determined rate. The plant inflow control valve is then operated by a device, which senses the level in the channel feeding the filters. The efficiency of turbidity removal in a filter is greatly reduced by sudden changes in flow. The total head loss across the filter and its control valve is equal to the difference between the water level in the filter chamber and the hydraulic grade level downstream of the control valve.

The flow is controlled by motion of the filter control valve, which automatically adjusts to compensate for changes in the resistance to flow in the filter bed and, because the filter bed clogs slowly, only a relatively slow movement is required. This allows the valve action to be well damped, so that 'hunting' will not occur, and sudden flow variations may be avoided.

With this control system, there is a sudden change of the flow through the pre-treatment portion of the plant when a filter is taken out of service for backwashing, and an even greater change as the filter chamber is refilled after backwashing. These sudden changes of flow can cause a marked deterioration in the settlement efficiency of any clarifier.

The system of upstream flow control with flow splitting avoids shock loading on the pre-treatment units. The plant inflow can either be set to a given flow rate or automatically controlled by the demand for water (perhaps from the level in the filter water storage). The flow to each filter is controlled by a flow-splitting device, through which the flow rate is a function of the level in the distribution channel. In this way, the flow from the pre-treatment section is split equally among all operating filters.

A filter that is relatively clean can filter water efficiently at a higher flow rate than one in which clogging with impurities is well advanced. A filter, which is starting to pass turbid water, can continue to produce acceptable water for a time, if the flow rate is reduced.

For some installations, it would appear that the best use could be made of a filter, if a high flow rate could be applied while it is relatively clean, with the rates of flow being gradually reduced as the filter becomes clogged. This is what happens if all filters are subjected to a common head loss. The total head loss through each filter, together with its underdrain and flow-limiting orifice, is equal to the difference in head between the influent header and the common effluent weir.

Filters are sometimes backwashed in a fixed sequence. The time for backwashing the next filter in the sequence is judged from the rising level of the water in all the filter basins. The improvement in performance, coupled with the absence of automatic control equipment, makes this filtration control ideal for use in countries with limited technology.

13.4 Hydraulics of filtration

The hydraulics of filtration are an important area for design and operation considerations. Accurate prediction of head loss and bed expansion during backwashing is important. Otherwise, filter design may prove to be inaccurate, and the sand in the bed may also be lost.

After the first backwash, rapid (gravity) sand filters will stratify (or be stratified by design in the case of multi-media filters). Each layer in the filter will consist of different sized particles.

After a filter has been in operation for a period of time, the head loss builds up as a result of filter clogging. When the head loss reaches an unacceptable level, the filter is stopped and backwashing is commenced. Unfortunately, backwashing of vertical-flow constructed wetlands is virtually impossible. However, the degradation of rhizomes results in hollow passages where liquid can rapidly pass through. This counteracts clogging, and makes backwashing unnecessary.

When a bed of uniform particles is subjected to backwashing, the bed just begins to open when the backwash velocity reaches a critical value. At this stage, the effective weight of the (submerged) particles is exactly balanced by the upward drag on the particles resulting from the upflow velocity. As the velocity increases, the bed opens up further (porosity increases). This increase in velocity does not improve the cleaning action, but is thought to be important in allowing sufficient open space for trapped suspended matter to be washed away.

For a stratified bed, expansion takes place successively for each media type. The surface strata are expanded at a lower rate of backwash than deeper ones. The bed is fully expanded when the upflow (backwashing) velocity of a filter equals the critical velocity for the largest particles.

13.5 Summary of rapid sand filtration

Rapid (gravity) sand filters operate at rates some ten times those of slow sand filters. It follows that impurities are drawn deep into the bed. Hence, cleaning is automated and hydraulic (i.e. not labour intensive). The rapid sand filter takes up a relatively small land area (in comparison to a slow sand filter or wetland), the water requires pre-treatment and the operator skill level required is high.

Chapter 14

Biological treatment

14.1 Aerobic self-purification

Essentially, (aerobic) biological treatment processes in wetland systems depend on supplying colonies of micro-organisms with optimum quantities of air (i.e. oxygen) and nutrients to achieve, at a maximum benefit to cost ratio, the same reactions that occur in natural self-purification processes. In general, three changes occur during self-purification (aerobic processes only). These are as follows:

- Coagulation of colloidal solids passing through the primary sedimentation stage;
- Oxidation of the following;

$$\text{Carbon:} \quad C \rightarrow CO_2, CO_3 \text{ and/or } HCO_3;$$
$$\text{Hydrogen:} \quad H \rightarrow H_2O;$$
$$\text{Nitrogen:} \quad N \rightarrow NH_2; \text{ and}$$
$$\text{Phosphorus:} \quad P \rightarrow PO_4.$$

- Nitrification: $\quad NH_3 \xrightarrow{\textit{Nitrobacter}} NO_2 \xrightarrow{\textit{Nitrosomonas}} NO_3.$

The basic requirements of any aerobic system for successful treatment of organic matter are as follows:

- A community of acclimatised micro-organisms;
- Adequate substrate (food); i.e. waste free from toxins; and
- A suitable environment; e.g., temperature, light, sufficient oxygen content and pH.

The basic biological treatment processes are as follows:

- Waste stabilisation ponds and constructed wetland systems (mixture of catabolic and anabolic processes);
- Trickling (or percolating) filter systems (primarily catabolic); and
- Activated sludge systems (primarily anabolic).

14.2 Waste stabilisation ponds

14.2.1 Aerobic ponds

In their simplest form, an aerobic stabilisation pond is a large and shallow excavation in the ground. The treatment of the waste is by natural processes involving the use of both bacteria and algae. These ponds can be aerobic (Scholz, 2003), anaerobic or facultative (both aerobic and anaerobic). The oxygen required in aerobic and facultative ponds is supplied by algae (through respiration during sunshine) and by diffusion from the atmosphere to the bulk liquid. Mixing in the ponds is usually achieved by wind (though some ponds are mechanically mixed).

In aerobic ponds, oxygen is supplied by natural surface re-aeration and by algal photosynthesis. Except for the algal population, the biological community (predominantly bacteria, protozoa and metazoa) in stabilisation ponds is similar to that in an activated sludge system. The oxygen released by the algae through the process of photosynthesis is used by the bacteria in the aerobic degradation of the organic matter. The nutrients and carbon dioxide released in this degradation process are (in turn) used by the algae.

Higher animals such as rotifers and protozoa are also present in the pond. Their main function is to predate on the bacteria, and to a lesser extent on algae. Therefore, higher animals control the suspended solids (SS) concentration in the effluent. This process is also called 'polishing'.

The presence of particular bacteria, algae and animals depends on variables such as the organic loading rate (kg BOD m^3/d), degree of pond mixing, pH, nutrients, sunlight and temperature. Temperature has a profound effect on the operation of aerobic ponds, particularly in regions with cold winters.

14.2.2 Facultative ponds

Ponds in which the stabilisation of wastes is brought about by a combination of aerobic, anaerobic and facultative bacteria are known as facultative stabilisation ponds. Three zones exist in such ponds, and can be described as follows:

- A surface zone where aerobic bacteria and algae exist in a symbiotic relationship;
- An intermediate zone that is partly aerobic and partly anaerobic (usually on a time basis) in which the decomposition of organic matter is carried out by facultative bacteria; and
- An anaerobic bottom zone in which accumulated solids are decomposed by anaerobic bacteria.

Conventional facultative ponds are excavations filled with screened and, in some cases, comminuted raw wastewater or primary effluent. They are approximately 1 m deep. The large solids settle out of solution to form an anaerobic sludge layer on the bottom of the pond. Soluble and colloidal organic material are oxidised by aerobic and facultative bacteria using the oxygen produced by the algae growing abundantly near the surface (within the photic depth). The carbon dioxide produced in the organic oxidation process serves as the carbon source for the algae. Anaerobic breakdown of the solids in the sludge layer results in the production of dissolved organics and gases such as CO_2, H_2S and CH_4, which are either oxidised by the aerobic bacteria or vented to the atmosphere.

In practice, oxygen is maintained in the upper layer of facultative ponds by the presence of algae and by surface re-aeration. The biological community in the upper (aerobic) layer is similar to that of an aerobic pond. The micro-organisms in the middle and bottom layers are facultative or strict anaerobes (usually these are only found in the bottom layer). Respiration (by bacteria) also occurs in the presence of sunlight. However, the net reaction in the pond is the production of oxygen. Because algae use carbon dioxide in photosynthetic activity, high pH conditions (pH > 10) can occur particularly in wastewater with low alkalinity (buffering capacity). The simplified reactions for both photosynthesis and respiration are presented in Eqs. (14.2.1) and (14.2.2).

$$\text{Photosynthesis:} \quad CO_2 + 2H_2O \xrightarrow{\text{in presence of light}} (CH_2O) + O_2 + H_2O$$

$$(14.2.1)$$

$$\text{Respiration:} \quad CH_2O + O_2 \rightarrow CO_2 + H_2O \qquad (14.2.2)$$

Facultative pond systems are natural systems, and as such require a minimum energy input in the form of electricity, but they do require large land areas and are difficult to control, particularly in temperate climates. These factors have led to the development of engineered systems such as the constructed wetland, percolating filter or the activated sludge process.

Chapter 15

Biological filtration

15.1 Introduction

The discussion is limited to aerobic processes only. However, it is important to remember that anaerobic biological filtration processes are used, particularly to treat strong industrial wastes, where the bulk of the biological oxygen demand (BOD) is present as soluble BOD.

Aerobic biological filtration processes have been around since the 1900s, though little was understood about their mode of operation until the early 1920s. Rational design methods for these units are still not in common use.

15.2 Trickling filter

The trickling (or percolating) filter, also called a bacteria bed, consists of a bed of suitable course porous media on which a biological film or slime grows. Hence, the general name for such types of processes is 'fixed film' system. The film consists mainly of bacteria upon which higher organisms graze. This grazing process helps to keep the film active.

Settled sewage is periodically distributed over the surface of the medium and, as it flows down through the bed, the fine suspended and dissolved organic matter is adsorbed by the bacteria. Oxygen to sustain aerobic biological oxidation is provided by air, which circulates through the bed via vents at the base of the filter.

Clogging of the interstices within the filter, sometimes termed 'ponding', is caused by excessive, unchecked and uncontrolled biological growth. This is normally prevented by the film or biomass being sheared off by the flow of wastewater. This material, which constitutes the net increase in biomass in the system and which would otherwise contribute to high BOD and SS concentrations in the effluent, is removed in secondary sedimentation tanks, called humus tanks.

It is difficult to obtain an adequate measure of the active mass of biological solids in a trickling filter. However, the total surface area of the medium gives some indication of the possible area available for growth and thus biomass. Indeed media manufacturers often quote the available surface area on the media. However, the actual thickness of the biofilm and the percentage of its active bacteria proportion cannot be practically determined without disturbing or even destroying at least some parts of the system.

Knowledge of the available surface area is often of little use to design engineers. Therefore, it is normal practice to use the volume of media as the measure of micro-organism activity in a biological filter; thus loading rates to biological filters are expressed in organic (kg BOD/m^3 of media/d) and/or volumetric (m^3/m^2 of surface area/d) units. The volumetric loading rate is important, since it affects the distribution of the flow over the surface of the medium, and hence the quality of the contact between the applied organic matter and the active biomass. Trickling filters can be compared to some types of constructed wetlands. However, they are not planted with macrophytes, but contain more voids to foster air entrainment.

15.3 Basic ecology

Trickling filters are well known for their diversity of lifeforms participating in the wastewater stabilisation. These include procaryotic and eucaryotic organisms as well as higher life forms such as rotifers, nematodes, annelid worms, snails and many insect larvae.

The bacteria are active in the uptake and degradation of soluble organic matter. Nitrifying bacteria convert ammonia to nitrate. In a low-rate trickling filter, there is a high nitrifier population and the effluent is well nitrified. In a high-rate filter, there is more sloughing of the biomass, due to higher fluid shear, and so little or no nitrification takes place.

Fungi are also to be found in trickling filters, these are also active in the biofilm and are actively involved in waste stabilisation. These organisms tend to dominate at lower pH values, usually associated with industrial waste treatment. Algae are once again active in the biofilm and produce oxygen during daylight hours. This helps to keep the uppermost portion of the biofilm aerobic. Both the fungi and algae are important components of the biofilm in trickling filters.

Protozoa are unicellular prokaryotic organisms, which feed on the bacteria within the biofilm. The continuous removal of bacteria by protozoa helps to maintain an active bacterial population, and thus maintains a high decomposition rate.

Rotifers are also present in the biofilm. Once again these organisms predate on bacteria, algae and fungi. Most rotifers are taken to be indicative of a high degree of treatment efficiency, and, when present, serve to reduce the effluent SS.

The major macroinvertebrates present in trickling filters are insect larvae (e.g., chironomids). These feed on the biofilm and help to control its thickness, thus avoiding clogging of the pores in the filter by microbial extra polymeric substances. The larvae develop into adult insects (filter flies) in two to three weeks. These can be a nuisance to plant operators and local residents, particularly in summer. Numbers as high as $3 \times 10^5/m^2$ have been reported. Insects are controlled by increasing the frequency of wetting of the filter surface (larvae only emerge on dry filters), by the use of insecticides or by biological control using *Bacillus thuringiensis* var. *israelensis*. This pathogen contains a toxin, which (when ingested by the insect) causes its death.

Other important controls on the ecology of the trickling filter include cold temperature and direct application of toxins. These slow down or stop predator activity, and thus increase the likelihood of clogging. This in turn may adversely affect the performance of the humus tank, particularly in spring when predator activity usually resumes after winter. Excess biofilm from the spring sloughing is likely to overload the humus tank, and causes high suspended solids to be discharged in the effluent. Nitrification is also reduced in winter as a result of the low average temperature.

15.4 Process variants

Various factors affect the efficiency of the aerobic biological filtration process; they include the following: organic loading rate; volumetric or hydraulic loading rate; depth of media; type, size and shape of media; and the total surface area of media available in the filter. A number of process variants have been developed over the years to deal with specific waste types and specific circumstances.

Standard rate filters have low hydraulic and organic loading rates, usually without recirculation; e.g., the effluent is applied to the filter, passes through, goes to the humus tank, and is then discharged. This type of filter will give an effluent with BOD <20 mg/l and SS <30 mg/l. Moreover, the effluent is well nitrified from a domestic sewage feedstock. Single pass units (i.e. through the filter once) can be fed by gravity without any source of external power. Typical loading rates are between 0.07 and 0.10 kg BOD/m³/d, and volumetric loading rates are between 0.25 and 1.2 m³/m²/d.

High-rate filters have loading rates, which are significantly higher than those of the low-rate filters. Recirculation is usually included and so pumping is normally necessary. Recirculation is a process whereby a proportion of the effluent is recycled to the influent of the filter and mixed with the incoming effluent, thus increasing the volume to be treated but decreasing its BOD.

Traditionally, rocks were used as filter media (aggregates). Nowadays, it is much more likely that the media consists of a plastic packing material. Organic rates are variable, and volumetric rates are up to 3.5 $m^3/m^2/d$.

High-rate filters are sometimes operated as part of a two-stage biological treatment process for strong effluents. In these cases, the organic loading rate can be up to 2 kg $BOD/m^3/d$ with volumetric loadings between 10 and 20 $m^3/m^2/d$, and with BOD removals between 60 and 65%. This type of filter is usually referred to as a roughing filter.

Alternating double filtration uses two filters in series. The effluent from the first filter passes to a humus tank from which the settled effluent is dosed onto a second filter and subsequently passes to a second humus tank. The first filter is subjected to a high organic loading rate, and so quickly accumulates a heavy biomass growth, while the second filter is lightly loaded. The filter sequence is altered on a weekly or monthly basis, depending on the degree of ponding present on the first filter.

When the flow pattern is changed, the previously lightly loaded filter begins to develop a greater biomass, while the previously highly loaded filter sheds biomass as the loading rate is now insufficient to sustain a similar growth rate. Effluents with BOD <20 mg/l and SS <30 mg/l are usually produced, though extensive pumping is required, and the design of the humus tanks must account for the increased solids loading.

15.5 Design of biological filters

Biological filters are suitable both as a complete secondary treatment unit or as a roughing process. Although a biological filter has a lower power requirement and thus a lower operational cost than the activated sludge process, it has a relatively high capital cost (particularly land requirement). This had made it unpopular for large treatment works. Moreover, biological filters can create a fly nuisance during summer. However, they do produce a well-nitrified effluent. The main aspects to be considered in designing these units are as follows:

- Pre-treatment requirements;
- Dosing system for applying effluent to the filter;
- Filter bed, volume and type;
- Dimensions of filter media including plastic material;
- Underdrain system;
- Ventilation system; and
- Humus tanks.

Pre-treatment is required to remove all material that is likely to clog the nozzles on the distribution arms. Preliminary and primary treatment are both used to achieve this. Moreover, they reduce the organic loading rate to the units.

Dosing equipment distributes the settled sewage regularly and evenly over the surface of the filter bed. A circular filter has a multiple arm rotary distributor supported on a central bearing, which incorporates the inlet. Rotation of the distributor is induced by the jet reaction of the flow from the nozzles along one side of the distributor arm. Occasionally, these arms require a brake to be fitted to slow down the speed of rotation. With a gravity-fed system, a dosing siphon and chamber are provided to hold back the flow until a sufficient volume is stored to ensure effective operation.

Filter beds are usually circular, though rectangular beds have been used in an effort to reduce land requirements (e.g., at Esholt, the sewage treatment works for Bradford, England). The media used in the filter must provide a suitable environment for the growth of the biofilm and additionally be inert to chemical attack.

The filter media for standard rate filters are usually gravel, crushed rock, blast furnace slag or other granular material. It is usual to limit the minimum size of particles between 40 and 50 mm, with typical sizes between 50 and 100 mm. The filter depth is of the order of 2 m. However, the actual volume of media required can be calculated.

The underdrain collects the effluent, and usually consists of an impervious base beneath the perforated filter floor, graded towards one or more collecting channels. The capacity of the underdrain should be large enough to accommodate the maximum flow rate whilst maintaining an adequate air space.

This is normally achieved by natural air circulation, induced by the heating or cooling effect of the applied wastewater. Sufficient openings should be provided at the base of the filter to allow air to flow into or out of the space under the suspended filter floor. In temperate climates, adequate ventilation can be assumed by providing plentiful air inlets around the base of the filter. Care must be taken in cold climates to minimise heat loss, and so to prevent low temperature inhibition of biological activity. In hot and dry climates, plant operators must prevent the air from drying out the biofilm, if the flow rate is low.

Furthermore, sloughing of biomass tends to be a seasonal phenomenon, occurring predominantly in spring. The corresponding humus sludge is difficult to dewater. However, constructed wetlands can be used for sludge dewatering in warm countries.

in particular, have made an important textbook contribution to constructed wetland research (Cooper et al., 1996). However, some of the information can be judged as being outdated.

16.2 Definitions

Defining wetlands has long been a problematic task, partly due to the diversity of environments, which are permanently or seasonally influenced by water, but also due to the specific requirements of the diverse groups of people involved with the study and management of these habitats. The Ramsar Convention, which brought wetlands to the attention of the international community, proposed the following definition (Convention on Wetlands of International Importance Especially as Waterfowl Habitat, 1971):

"Wetlands are areas of marsh, fen, peatland or water, whether natural or artificial, permanent or temporary, with water that is static or flowing, fresh, brackish or salt, including areas of marine water, the depth of which at low tide does not exceed six metres".

Another, more succinct, definition is as follows (Smith, 1980):

"Wetlands are a half-way world between terrestrial and aquatic ecosystems and exhibit some of the characteristics of each".

This complements the Ramsar description, since wetlands are at the interface between water and land. This concept is particularly important in areas where wetlands may only be 'wet' for relatively short periods of time during a year, such as in areas of the tropics with marked wet and dry seasons.

These definitions put an emphasis on the ecological importance of wetlands. However, the natural water purification processes occurring within these systems have become increasingly relevant to those people involved with the practical use of constructed or even semi-natural wetlands for water and wastewater treatment. There is no single accepted ecological definition of wetlands.

Wetlands are characterised by the following (United States Army Corps of Engineers, 2000):

- The presence of water;
- Unique soils that differ from upland soils; and
- The presence of vegetation adapted to saturated conditions.

Whichever definition is adopted, it can be seen that wetlands encompass a wide range of hydrological and ecological types, from high altitude river sources (contradiction to United States Army Corps of Engineers (2000)), to shallow coastal regions, in each case being affected by prevailing climatic conditions. For the purpose of this chapter, however, the main emphasis will be upon constructed treatment wetlands in a temperate climate.

16.3 Hydrology of wetlands

The biotic status of a wetland is intrinsically linked to the hydrological factors by which it is affected. These affect the nutrient availability as well as physico-chemical variables such as soil and water pH, and anaerobiosis within soils. In turn, biotic processes will have an impact upon the hydrological conditions of a wetland.

Water is the hallmark of wetlands. Therefore, it is not surprising that the input and output of water (i.e. water budget; see below) of these systems determine the biochemical processes occurring within them. The net result of the water budget (hydroperiod) may show great seasonal variations, but ultimately delineates wetlands from terrestrial and fully aquatic ecosystems.

From an ecological point of view, as well as an engineering one, the importance of hydrology cannot be overstated, as it defines the species diversity, productivity and nutrient cycling of specific wetlands. That is to say, hydrological conditions must be considered, if one is interested in the species richness of flora and fauna, or if the interest lies in utilising wetlands for pollution control.

16.3.1 Hydroperiod and water budget

The stability of particular wetlands is directly related to their hydroperiod; that is the seasonal shift in surface and sub-surface water levels. The terms flood duration and flood frequency refer to wetlands that are not permanently flooded and give some indication of the time period involved in which the effects of inundation and soil saturation will be most pronounced.

Of particular relevance to riparian wetlands is the concept of flooding pulses as described by Junk et al. (1989). These pulses cause the greatest difference in high and low water levels, and benefit wetlands by the inflow of nutrients and washing out of waste matter. These sudden and high volumes of water can be observed on a periodic or seasonal basis. It is particularly important to appreciate this natural fluctuation and its effects, since wetland management often attempts to control the level by which waters rise and fall. Such manipulation might be due to the overemphasis placed on water and its role in the lifecycles of wetland flora and fauna, without considering the fact that such species have evolved in such an unstable environment (Fredrickson and Reid, 1990).

The balance between the input and output of water within a wetland is called 'water budget'. This budget is summarised by Eq. (16.3.1), where the volumetric units could be m^3.

$$\Delta V / \Delta t = P_n + S_i + G_i - ET - S_o - G_o \pm T \qquad (16.3.1)$$

where,

V = volume of water storage within a wetland;
$\Delta V / \Delta t$ = change in volume of water storage in a wetland per unit time (t);
P_n = net precipitation;
S_i = surface inflows including flooded streams;
G_i = groundwater inflows;
ET = evapotranspiration;
S_o = surface outflows;
G_o = groundwater outflows; and
T = tidal inflow (+) or outflow (−).

16.3.2 Precipitation, interception, through-fall and stem-flow

In general terms, wetlands are most widespread in those parts of the world where precipitation exceeds water loss through evapotranspiration and surface runoff. The contribution of precipitation to the hydrology of a wetland is influenced by a number of factors. Precipitation such as rain and snow often passes through a canopy of vegetation before it becomes part of the wetland. The volume of water retained by this canopy is termed interception. Variables such as precipitation intensity and vegetation type will affect interception, for which median values of several studies have been calculated as 13% for deciduous forests and 28% for coniferous woodland (Dunne and Leopold, 1978).

The precipitation that remains to reach the wetland is termed the through-fall. This is added to the stem-flow, which is the water running down vegetation stems and trunks, generally considered a minor component of a wetland water budget, such as 3% of through-fall in cypress dome wetlands in Florida (Heimburg, 1984). Thus through-fall and stem-flow form Eq. (16.3.2), where the volumetric units could be m^3; the most commonly used precipitation equation for wetlands.

$$P_n = TF + SF \qquad (16.3.2)$$

where,

P_n = net precipitation;
TF = through-fall; and
SF = stem-flow.

16.4 Wetland chemistry

16.4.1 Oxygen

Because wetlands are associated with waterlogged soils, the concentration of oxygen within sediments and the overlying water is of critical importance. The rate of oxygen diffusion into water and sediment is slow, and this (coupled with microbial and animal respiration) leads to near anaerobic sediments within many wetlands (Moss, 1998). These conditions favour rapid peat build up, since decomposition rates and inorganic content of soils are low. Furthermore, the lack of oxygen in such conditions affects the aerobic respiration of plant roots and influences plant nutrient availability. Wetland plants have consequently evolved to be able to exist in anaerobic soils.

While the deeper sediments are generally anoxic, a thin layer of oxidised soil usually exists at the soil–water interface. The oxidised layer is important, since it permits the oxidised forms of prevailing ions to exist. This is in contrast to the reduced forms occurring at deeper levels of soil. The state of reduction or oxidation of iron, manganese, nitrogen and phosphorus ions determines their role in nutrient availability and also toxicity. The presence of oxidised ferric iron (Fe^{3+}) gives the overlying wetland soil a brown coloration, whilst reduced sediments have undergone 'gleying', a process by which ferrous iron (Fe^{2+}) gives the sediment a blue-grey tint.

Therefore, the level of reduction of wetland soils is important in understanding the chemical processes that are most likely to occur in the sediment and influence the above water column. The most practical way to determine the reduction state is by measuring the redox potential, also called the oxidation–reduction potential, of the saturated soil or water. The redox potential quantitatively determines whether a soil or water sample is associated with a reducing or oxidising environment. Reduction is the release of oxygen and gain of an electron (or hydrogen), while oxidation is the reverse; i.e. the gain of oxygen and loss of an electron. This is shown by Eq. (16.4.3), and explained in detail by Mitsch and Gosselink (2000).

$$E_H = E^0 + 2.3[RT/nF] \log[\{ox\}/\{red\}] \qquad (16.4.3)$$

where,

E_H = redox potential (hydrogen ion scale);
E^0 = potential of reference (mV);
R = gas constant = 81.987 cal/deg/mol;
T = temperature (°C);
n = number of moles of electrons transferred; and
F = Faraday constant = 23 061 cal/mole/volt.

Oxidation (and therefore decomposition) of organic matter (a very reduced material) occurs in the presence of any electron acceptor, particularly oxygen, although NO^{3-}, Mn^{2+}, Fe^{3+} and SO_4^{2-} are also commonly involved in oxidation, but the rate will be slower in comparison with oxygen. A redox potential range between +400 mV and +700 mV is typical for environmental conditions associated with free dissolved oxygen. Below +400 mV, the oxygen concentration will begin to diminish and wetland conditions will become increasingly more reduced (>−400 mV).

Redox potentials are affected by pH and temperature, which influences the range at which particular reactions occur. The following thresholds are therefore not definitive:

- Once wetland soils become anaerobic, the primary reaction at approximately +250 mV is the reduction of nitrate (NO_3^-) to nitrite (NO_2^-), and finally to nitrous oxide (N_2O) or free nitrogen gas (N_2).
- At about +225 mV, manganese is reduced to manganous compounds. Under further reduced conditions, ferric iron becomes ferrous iron between approximately +100 mV and −100 mV, and sulphates become sulphides between approximately −100 and −200 mV.
- Under the most reduced conditions (<−200 mV) the organic matter itself and/or carbon dioxide will become the terminal electron acceptor. This results in the formation of low molecular weight organic compounds and methane gas ($CH_4 \uparrow$).

16.4.2 Carbon

Organic matter within wetlands is usually degraded by aerobic respiration or anaerobic processes (e.g., fermentation and methanogenesis). Anaerobic degradation of organic matter is less efficient than decomposition occurring under aerobic conditions.

Fermentation is the result of organic matter acting as the terminal electron acceptor (instead of oxygen as in aerobic respiration). This process forms low molecular weight acids (e.g., lactic acid), alcohols (e.g., ethanol) and carbon dioxide. Therefore, fermentation is often central in providing further biodegradable substrates for other anaerobic organisms in waterlogged sediments.

The sulphur cycle is linked with the oxidation of organic carbon in some wetlands, particularly in sulphur-rich coastal systems. Low-molecular weight organic compounds that result from fermentation (e.g., ethanol) are utilised as organic substrates by sulphur-reducing bacteria during the conversion of sulphate to sulphide (Mitsch and Gosselink, 2000).

Previous work suggests that methanogenesis is the principal carbon pathway in freshwater. Between 30 and 50% of the total benthic carbon flux has been attributed to methanogenesis (Boon and Mitchell, 1995).

16.4.3 Nitrogen

The prevalence of anoxic conditions in most wetlands has led to them playing a particularly important role in the release of gaseous nitrogen from the lithosphere and hydrosphere to the atmosphere through denitrification (Mitsch and Gosselink, 2000). However, the various oxidation states of nitrogen within wetlands are also important to the biogeochemistry of these environments.

Nitrates are important terminal electron acceptors after oxygen, making them relevant in the process of oxidation of organic matter. The transformation of nitrogen within wetlands is strongly associated with bacterial action. The activity of particular bacterial groups is dependent on whether the corresponding zone within a wetland is aerobic or anaerobic.

Within flooded wetland soils, mineralised nitrogen occurs primarily as ammonium (NH_4^+), which is formed through ammonification, the process by which organically bound nitrogen is converted to ammonium nitrogen under aerobic or anaerobic conditions. Soil-bound ammonium can be absorbed through rhizome and root systems of macrophytes and be reconverted to organic matter, a process that can also be performed by anaerobic micro-organisms.

The oxidised top layer present in many wetland sediments, is crucial in preventing the excessive build-up of ammonium. A concentration gradient will be established between the high concentration of ammonium in the lower reduced sediments and the low concentration in the oxidised top layer. This may cause a passive flow of ammonium from the anaerobic to the aerobic layer, where microbiological processes convert the ion into further forms of nitrogen.

Within the aerobic sediment layer, nitrification of ammonium, firstly to nitrite (NO_2^-) and subsequently to nitrate (NO_3^-), is shown in the Eqs. (16.4.4) and (16.4.5), preceded by the genus of bacteria predominantly involved in each process step. Nitrification may also take place in the oxidised rhizosphere of wetland plants.

$$\textit{Nitrosomonas: } 2NH_4^+ + 3O_2 \rightarrow 2NO_2^- + 2H_2O + 4H^+ + \text{energy} \quad (16.4.4)$$

$$\textit{Nitrobacter: } 2NO_2^- + O_2 \rightarrow 2NO_3^- + \text{energy} \quad (16.4.5)$$

A study in southern California indicated that denitrification was the most likely pathway for nitrate loss from experimental macrocosms, and larger constructed wetlands (Bachand and Horne, 1999). Very high rates of nitrate-nitrogen removal were reported (2800 mg $N/m^2/d$). Furthermore, nitrate removal from inflow (waste)water is generally lower in constructed wetlands compared to natural systems (Spieles and Mitsch, 2000). There is considerable interest in enhancing bacterial denitrification in constructed wetlands in order to reduce the level of eutrophication in receiving waters such as rivers and lakes (Bachand and Horne, 1999).

An investigation into the seasonal variation of nitrate removal showed maximum efficiency to occur during summer. This study also indicated a seasonal relationship in the pattern of nitrate retention, in which nitrate assimilation and denitrification are temperature dependent (Spieles and Mitsch, 2000).

Further evidence supporting the importance of denitrification was presented by Lund et al. (2000). The proportion of nitrogen removed by denitrification from a wetland in southern California was estimated by analysing the increase in the proportion of the nitrogen isotope ^{15}N found in the outflow water. This method is based on the tendency of the lighter isotope ^{14}N to be favoured by the biochemical thermodynamics of denitrification, thus reducing its proportion in water flowing out of wetlands in which denitrification is prevalent. Denitrification seems to be the favoured pathway of nitrate loss from a treatment wetland, as this permanently removes nitrogen from the system, compared to sequestration within algal and macrophyte biomass.

In some wetlands, nitrogen may be derived through nitrogen fixation. In the presence of the enzyme nitrogenase, nitrogen gas is converted to organic nitrogen by organisms such as aerobic or anaerobic bacteria and cyanobacteria (blue-green algae). Wetland nitrogen fixation can occur in the anaerobic or aerobic soil layer, overlying water, rhizosphere of plant roots, and on leaf or stem surfaces. Cyanobacteria may contribute significantly to nitrogen fixation.

In northern bogs, which are often too acidic for large bacterial populations, nitrogen fixation by cyanobacteria is particularly important (Etherington, 1983). However, it should be noted that while cyanobacteria are adaptable organisms, they are affected by environmental stresses. For example, cyanobacteria are particularly susceptible to ultraviolet radiation, whereby their nitrogen metabolism (along with other functions) is impaired (Donkor and Häder, 1996).

16.4.4 Phosphorus

In wetland soils, phosphorus occurs as soluble or insoluble, organic or inorganic complexes. Its cycle is sedimentary rather than gaseous (as with nitrogen) and predominantly forms complexes within organic matter in peatlands or inorganic sediments in mineral soil wetlands. Over 90% of the phosphorus load in streams and rivers may be present in particulate inorganic form (Overbeck, 1988).

Soluble reactive phosphorus is the analytical term given to biologically available ortho-phosphate, which is the primary inorganic form. The availability of phosphorus to plants and micro-consumers is limited due to the following main effects:

- Under aerobic conditions, insoluble phosphates are precipitated with ferric iron, calcium and aluminium;

- Phosphates are adsorbed onto clay particles, organic peat, and ferric and aluminium hydroxides and oxides; and
- Phosphorus is bound up in organic matter through incorporation into bacteria, algae and vascular macrophytes.

There are three general conclusions about the tendency of phosphorus to precipitate with selected ions (Reddy, 1999):

1. In acid soils, phosphorus is fixed as aluminium and iron phosphates;
2. In alkaline soils, phosphorus is bound by calcium and magnesium; and
3. The bioavailability of phosphorus is greatest at neutral to slightly acid pH.

The phosphorus availability is altered under anaerobic wetland soil conditions. The reducing conditions that are typical of flooded soils do not directly affect phosphorus. However, the association of phosphorus with other elements that undergo reduction has an indirect effect upon the phosphorus in the environment. For example, as ferric iron is reduced to the more soluble ferrous form, phosphorus as ferric phosphate is released into solution (Faulkner and Richardson, 1989; Gambrell and Patrick, 1978). Phosphorus may also be released into solution by a pH change brought about by organic, nitric or sulphuric acids produced by chemosynthetic bacteria. Phosphorus sorption to clay particles is greatest under strongly acidic to slightly acidic conditions (Stumm and Morgan, 1996).

Great temporal variability in phosphorus concentrations of wetland influent in Ohio, for example, has been reported (Nairn and Mitsch, 2000). However, no seasonal pattern in phosphorus concentration was observed. This was explained by precipitation events and river flow conditions. Dissolved reactive phosphorus levels peaked during floods and on isolated occasions in late autumn. Furthermore, sedimentation of suspended solids appears to be important in phosphorus retention within wetlands (Fennessy et al., 1994).

The physical, chemical and biological characteristics of a wetland system affect the solubility and reactivity of different forms of phosphorus. Phosphate solubility is regulated by temperature (Holdren and Armstrong, 1980), pH (Mayer and Kramer, 1986), redox potential (Moore and Reddy, 1994), the interstitial soluble phosphorus level (Kamp-Nielson, 1974) and microbial activity (Gächter and Meyer, 1993; Gächter et al., 1988).

Where agricultural land has been converted to wetlands, there can be a tendency in solubilisation of residual fertiliser phosphorus, which results in a rise of the soluble phosphorus concentration in floodwater. This effect can be reduced by physicochemical amendment, applying chemicals such as alum and calcium carbonate to stabilise the phosphorus in the sediment of these new wetlands (Ann et al., 1999a,b).

The redox potential has a significant impact on dissolved reactive phosphorus of chemically amended soils (Ann et al., 1999a,b). The redox potential can alter with fluctuating watertable levels and hydraulic loading rates. Dissolved phosphorus concentrations are relatively high under reduced conditions, and decrease with increasing redox potential. Iron compounds (e.g., $FeCl_3$) are particularly sensitive to the redox potential, resulting in the chemical amendment of wetland soils. Furthermore, alum and calcium carbonate are suitable to bind phosphorus even during fluctuating redox potentials.

Macrophytes assimilate phosphorus predominantly from deep sediments, thereby acting as nutrient pumps (Carignan and Kaill, 1980; Mitsch and Gosselink, 2000; Smith and Adams, 1986). The most important phosphorus retention pathway in wetlands is via physical sedimentation (Wang and Mitsch, 2000).

Model simulations on constructed wetlands in north-eastern Illinois (USA), for example, showed an increase in total phosphorus in the water column in the presence of macrophytes mainly during the non-growing period, with little effect during the growing season. Most phosphorus taken from sediments by macrophytes is reincorporated into the sediment as dead plant material, and therefore remains in the wetland indefinitely. Macrophytes can be harvested as a means to enhance phosphorus removal in wetlands. By harvesting macrophytes at the end of the growing season, phosphorus can be removed from the internal nutrient cycle within wetlands (Wang and Mitsch, 2000).

Moreover, the model showed a phosphorus removal potential of three-quarters of that of the phosphorus inflow. Therefore, harvesting would reduce phosphorus levels in upper sediment layers and drive phosphorus movement into deeper layers, particularly the root zone. In deep layers of sediment, the phosphorus sorption capacity increases along with a lower desorption rate (Wang and Mitsch, 2000).

16.4.5 Sulphur

In wetlands, sulphur is transformed by microbiological processes and occurs in several oxidation stages. Reduction may occur, if the redox potential is between -100 and -200 mV. Sulphides provide the characteristic 'bad egg' odour of wetland soils.

Assimilatory sulphate reduction is accomplished by obligate anaerobes such as *Desulfovibrio* spp. Bacteria may use sulphates as terminal electron acceptors (Eq. (16.4.6)) in anaerobic respiration at a wide pH range, but highest around neutral (Mitsch and Gosselink, 2000).

$$4H_2 + SO_4^{2-} \rightarrow H_2S \uparrow + 2H_2O + 2OH^- \tag{16.4.6}$$

The greatest loss of sulphur from freshwater wetland systems to the atmosphere is via hydrogen sulphide ($H_2S\uparrow$). In oceans, however, this is through the production of dimethyl sulphide from decomposing phytoplankton (Schlesinger, 1991).

Oxidation of sulphides to elemental sulphur and sulphates can occur in the aerobic layer of some soils and is carried out by chemoautotrophic (e.g., *Thiobacillus* spp.) and photosynthetic micro-organisms. *Thiobacillus* spp. may gain energy from the oxidation of hydrogen sulphide to sulphur, and further, by certain other species of the genus, from sulphur to sulphate (Eqs. (16.4.7) and (16.4.8)).

$$2H_2S + O_2 \rightarrow 2S + 2H_2O + \text{energy} \tag{16.4.7}$$

$$2S + 3O_2 + 2H_2O \rightarrow 2H_2SO_4 + \text{energy} \tag{16.4.8}$$

In the presence of light, photosynthetic bacteria, such as purple sulphur bacteria of salt marshes and mud flats, produce organic matter as indicated in Eq. (16.4.9). This is similar to the familiar photosynthesis process, except that hydrogen sulphide is used as the electron donor instead of water.

$$CO_2 + H_2S + \text{light} \rightarrow CH_2O + S \tag{16.4.9}$$

Direct toxicity of free sulphide in contact with plant roots has been noted. There is a reduced toxicity and availability of sulphur for plant growth, if it precipitates with trace metals. For example, the immobilisation of zinc and copper by sulphide precipitation is well known.

The input of sulphates to freshwater wetlands, in the form of Aeolian dust or as anthropogenic acid rain, can be significant. Sulphate deposited on wetland soils may undergo dissimilatory sulphate reduction by reaction with organic substrates (Eq. (16.4.10)).

$$2CH_2O + SO_4^{2-} + H^+ \rightarrow 2CO_2 + HS^- + 2H_2O \tag{16.4.10}$$

Protons consumed during this reaction (Eq. (16.4.10)) generate alkalinity. This is illustrated by the increase in pH with depth in wetland sediments (Morgan and Mandernack, 1996). It has been suggested that this 'alkalinity effect' can act as a buffer in acid rain affected lakes and streams (Rudd et al., 1986; Spratt and Morgan, 1990).

The sulphur cycle can vary greatly within different zones of a particular wetland. The stable isotope $\delta^{34}S$ within peat, the $^{35}SO_4^{2-}$:Cl^- ratio and the stable isotopes $\delta^{18}O$ and $\delta^{34}S$ of sulphate within different waters were analysed by Mandernack et al. (2000). The variability in the sulphur cycle within the watershed can affect the distribution of reduced sulphur stored in soil. This change in local sulphur availability can have marked effects upon stream water over short distances.

The estimation of generated alkalinity may be complicated due to the potential problems associated with the use of the $^{35}SO_4^{2-}$:Cl^- ratio and/or the $\delta^{34}S$ value, which are used to estimate the net sulphur retention. These problems may exist because ester sulphate pools can be a source of sulphate availability for sulphate reduction, and as a $\delta^{34}S$ sulphate buffer within stream water.

16.5 Wetland ecosystem mass balance

The general mass balance for a wetland system, in terms of chemical pathways, uses the following main pathways: inflows, intrasystem cycling and outflows. The inflows are mainly through hydrologic pathways such as precipitation, (particularly urban) surface runoff and groundwater. The photosynthetic fixation of both atmospheric carbon and nitrogen are important biological pathways. Intrasystem cycling is the movement of chemicals in standing stocks within wetlands, such as litter production and remineralisation. Translocation of minerals within plants is an example of the physical movement of chemicals. Outflows involve hydrologic pathways, but also include the loss of chemicals to deeper sediment layers, beyond the influence of internal cycling (although the depth at which this threshold occurs is not certain). Furthermore, the nitrogen cycle plays an important role in outflows, such as nitrogen gas lost as a result of denitrification. However, respiratory loss of carbon is also an important biotic outflow.

There is great variation in the chemical balance from one wetland to another, but the following generalisations can be made:

- Wetlands act as sources, sinks or transformers of chemicals depending on wetland type, hydrological conditions and length of time the wetland has received chemical inputs. As sinks, the long-term sustainability of this function is associated with hydrologic and geomorphic conditions as well as the spatial and temporal distribution of chemicals within wetlands.
- Particularly in temperate climates, seasonal variation in nutrient uptake and release is expected. Chemical retention will be greatest in the growing seasons (spring and summer) due to higher rates of microbial activity and macrophyte productivity.
- The ecosystems connected to wetlands affect and are affected by the adjacent wetland. Upstream ecosystems are sources of chemicals, whilst those downstream may benefit from the export of certain nutrients or the retention of particular chemicals.
- Nutrient cycling in wetlands differs from that in terrestrial and aquatic systems. More nutrients are associated with wetland sediments than with most terrestrial soils, whilst benthic aquatic systems have autotrophic activity, which relies more on nutrients in the water column than in the sediments.
- The ability of wetland systems to remove anthropogenic waste is not limitless.

Equation (16.5.11) indicates a general mass balance for a pollutant within a treatment wetland. Within this equation, transformations and accretion are long-term sustainable removal processes, whilst storage does not result in long-term average removal, but can lessen or accentuate the cyclic activity.

$$In - Out = Transformation + Accretion + Biomass\ Storage$$
$$+ Water/Soil\ Storage \qquad (16.5.11)$$

16.6 Macrophytes in wetlands

Wetland plants are often central to wastewater treatment wetlands. The following requirements of plants should be considered for use in such systems (Tanner, 1996):

- Ecological adaptability (no disease or weed risk to the surrounding natural ecosystems);
- Tolerance of local conditions in terms of climate, pests and diseases;
- Tolerance of pollutants and hypertrophic waterlogged conditions;
- Ready propagation, rapid establishment, spread and growth; and
- High pollutant removal capacity, through direct assimilation or indirect enhancement of nitrification, denitrification and other microbial processes.

Interest in macrophyte systems for sewage treatment by the UK water industry dates back to 1985 (Parr, 1990). The ability of different macrophyte species and their assemblages within systems to most efficiently treat wastewater has been examined previously (Kuehn and Moore, 1995). The dominant species of macrophyte varies from locality to locality. The number of genera (e.g., *Phragmites* spp., *Typha* spp. and *Scirpus* spp.) common to all temperate locations is great.

The improvement of water quality with respect to key water quality variables including BOD, COD, total SS, nitrates and phosphates has been studied by Turner (1995). Relatively little work has been conducted on the enteric bacteria removal capability of macrophyte systems (Perkins and Hunter, 2000).

16.6.1 Primary productivity

There have been many studies to determine the primary productivity of wetland macrophytes, although estimates have generally tended to be fairly high (Mitsch and Gosselink, 2000). The estimated dry mass production for *Phragmites australis* (Cav.) Trin. ex Steud. (common reed) is between 1000 and 6000 $g/m^2/a$ in the Czech Republic (Kvet and Husak, 1978), between 2040 and 2210 $g/m^2/a$ for *Typha latifolia* L. (cattail) in Oregon, USA (McNaughton, 1966), and 943 $g/m^2/a$ for *Scirpus fluviatilis* (Torr.) A. Gray [JPM][H&C] (river bulrush) in Iowa, USA (van der Valk and Davis, 1978).

Little of aquatic plant biomass is consumed as live tissue; it rather enters the pool of particulate organic matter following tissue death. The breakdown of this material is consequently an important process in wetlands and other shallow aquatic habitats (Gessner, 2000). Litter breakdown has been studied along with intensive work on *P. australis*, one of the most widespread aquatic macrophyte (Wrubleski et al., 1997).

There has been an emphasis on studying the breakdown of aquatic macrophytes in such a way, which most closely resembles that of natural plant death and decomposition, principally by not removing plant tissue from macrophyte stands. Many species of freshwater plants exhibit the so-called 'standing-dead' decay, which describes the observation of leaves remaining attached to their stems after senescence and death (Kuehn et al., 1999). Different fractions (leaf blades, leaf sheaths and culms) of *P. australis* differ greatly in structure and chemical composition, and may exhibit different breakdown rates, patterns and nutrient dynamics (Gessner, 2000).

16.6.2 *Phragmites australis*

Phragmites australis (Cav.) Trin. ex Steud. (common reed), formerly known as *Phragmites communis* (Norfolk reed), is a member of the large family Poaceae (roughly 8000 species within 785 genera). Common reed occurs throughout Europe to 70° North and is distributed worldwide. It may be found in permanently flooded soils of still or slowly flowing water. This emergent plant is usually firmly rooted in wet sediment but may form lightly anchored rafts of 'hover reed'.

It tends to be replaced by other species at drier sites. The density of this macrophyte is reduced by grazing (e.g., consumption by waterfowl) and may then be replaced by other emergent species such as *Phalaris arundinacea* L. (reed canary grass).

Phragmites australis is a perennial, and its shoots emerge in spring. Hard frost kills these shoots, illustrating the tendency for reduced vigour towards the northern end of its distribution. The hollow stems of the dead shoots in winter are important in transporting oxygen to the relatively deep rhizosphere (Brix, 1989).

Reproduction in closed stands of this species is mainly by vegetative spread, although seed germination enables the colonisation of open habitats. Detached shoots often survive, and regenerate away from the main stand (Preston and Croft, 1997).

Common reed or Norfolk Reed is most frequently found in nutrient-rich sites, and absent from the most oligotrophic zones. However, the stems of this species may be weakened by nitrogen-rich water, and are subsequently more prone to wind and wave damage, leading to an apparent reduction in density of this species in Norfolk (England), and elsewhere in Europe (Boar et al., 1989; Ostendorp, 1989).

16.6.3 *Typha latifolia*

Typha latifolia L. (cattail, reedmace or bulrush) is a species belonging to the small family Typhaceae. This species is widespread in temperate parts of the northern hemisphere but extends to South Africa, Madagascar, Central America and the West Indies, and has been naturalised in Australia (Preston and Croft, 1997). *Typha latifolia* is typically found in shallow water or on exposed mud at the edge of lakes, ponds, canals and ditches, and less frequently near fast flowing water. This species rarely grows at water depths >0.3 m, where it is frequently replaced by *P. australis*.

Reedmace is a shallow-rooted perennial producing shoots throughout the growing season, which subsequently die in autumn. Colonies of this species expand by rhizomatous growth at rates of 4 m/a, whilst detached portions of rhizome may float and establish new colonies. In contrast, colony growth by seeds is less likely. Seeds require moisture, light and relatively high temperatures to germinate, although this may occur in anaerobic conditions. Where light intensity is low, germination is stimulated by temperature fluctuation (Hammer, 1989).

16.7 Physical and biochemical parameters

The key physicochemical parameters relevant for wetland systems include the BOD, turbidity and the redox potential. The BOD is an empirical test to determine the molecular oxygen used during a specified incubation period (usually five days), for the biochemical degradation of organic matter (carbonaceous demand) and the oxygen used to oxidise inorganic matter (e.g., sulphides and ferrous iron). An extended test (up to 25 days) may also measure the amount of oxygen used to oxidise reduced forms of nitrogen (nitrogenous demand), unless this is prevented by an inhibitor chemical (Scholz, 2004a). Inhibiting the nitrogenous oxygen demand is recommended for secondary effluent samples (Clesceri et al., 1998).

The European Union freshwater fisheries directive sets an upper BOD limit of 3 mg/l for salmonid rivers and 6 mg/l for coarse fisheries. A river is deemed polluted, if the BOD exceeds 5 mg/l. Municipal wastewater values are usually between approximately 150 and 1000 mg/l (Kiely, 1997).

Turbidity is a measure of the cloudiness of water, caused predominantly by suspended material such as clay, silt, organic and inorganic matter, plankton and other microscopic organisms, and scattering and absorbing light. Turbidity in wetlands and lakes is often due to colloidal or fine suspensions, whilst in fast-flowing waters the particles are larger and turbid conditions are prevalent predominantly during floods (Kiely, 1997).

The redox potential is another key parameter for monitoring wetlands. The reactivities and mobilities of elements such as Fe, S, N, C and a number of metallic elements depend strongly on the redox potential conditions. Reactions involving electrons and protons are pH- and redox-potential-dependent. Chemical reactions in aqueous media can often be characterised by pH and the redox potential together with the activity of dissolved chemical species. The redox potential is a measure of intensity, and does not represent the capacity of the system for oxidation or reduction (American Public Health Association, 1995). The interpretation of the redox potential values measured in the field is limited by a number of factors, including irreversible reactions, 'electrode poisoning' and multiple redox couples.

16.8 Examples for natural and constructed wetlands

16.8.1 Riparian wetlands

Riparian wetlands are ecosystems under the influence of adjacent streams or rivers (Scholz and Trepel, 2004). A succinct definition is as follows (Gregory et al., 1991):

> "Riparian zones are the interface between terrestrial and aquatic ecosystems. As ecotones, they encompass sharp gradients of environmental factors, ecological processes and plant communities. Riparian zones are not easily delineated, but are composed of mosaics of landforms, communities and environments within the larger landscape".

There are four main reasons as to why periodic flooding, which is typical of riparian wetlands, contributes to the observed higher productivity compared to adjacent upland ecosystems:

(1) There is an adequate water supply for vegetation.
(2) Nutrients are supplied and coupled with a favourable change in soil chemistry (e.g., nitrification, sulphate reduction and nutrient mineralisation).
(3) In comparison to stagnant water conditions, a more oxygenated root zone follows flooding.
(4) Waste products (e.g., carbon dioxide and methane) are removed by periodic 'flushing'.

Nutrient cycles in riparian wetlands can be described as follows:

• Nutrient cycles are 'open' due to the effect of river flooding, runoff from upslope environments or both (depending on season and inflow stream, or river type).

- Riparian forests have a great effect on the biotic interactions within intra-system nutrient cycles. The seasonal pattern of growth and decay often matches available nutrients.
- Water in contact with the forest floor leads to important nutrient transformations. Therefore, riparian wetlands can act as sinks for nutrients that enter, for example, as urban runoff and groundwater flow.
- Riparian wetlands have often appeared to be nutrient transformers, changing a net input of inorganic nutrients to a net output of their corresponding organic forms.

The nitrogen cycle within a temperate stream-floodplain environment is of particular interest to ecological engineers. During winter, flooding contributes to the accumulation of dissolved and particulate organic nitrogen that is not assimilated by the trees due to their dormancy. This fraction of nitrogen is retained by filamentous algae and through immobilisation by detritivores on the forest floor.

As the water of a wetland gets warmer in spring, nitrogen is released by decomposition and by shading of the filamentous algae by the developing tree canopy. Nitrate is then assumed to be immobilised in the decaying litter and gradually made available to plants. As vegetation increases, the plants take up more nitrogen, and water levels fall subsequently due to evapotranspiration. Ammonification and nitrification rates increase with exposure of the sediments to the atmosphere. Nitrates produced in nitrification are lost when denitrification becomes prevalent as flooding later in the year creates anaerobic conditions.

In terms of reducing the effects of eutrophication on open water by urban runoff, the use of riparian buffer zones, particularly of *Alnus incana* (grey alder) and *Salix* spp. (willow) in conjunction with perennial grasses has been recommended (Mander et al., 1995). Riparian zones are also termed riparian forest buffer systems (Lowrance et al., 1979). Such zones were found to reduce the nutrient flux into streams.

The role of riparian ecosystems in nutrient transformations is specifically important in relation to the production of the greenhouse gas nitrous oxide (N_2O). Due to the inflow of excess agricultural nitrogen into wetland systems, the riparian zones, in particular, are likely 'hot spots' for nitrous oxide production (Groffman, 2000).

The control of non-point source pollution can be successfully achieved by riparian forest buffers in some agricultural watersheds and most effectively, if excess precipitation moves across, in or near the root zone of the riparian forest buffers. For example, between 50 and 90% retention of total nitrate loading in both shallow groundwater and sediment subject to surface runoff within the Chesapeake Bay watershed (USA) was observed. In comparison, phosphorus retention was found to be generally less than nitrate retention (Lowrance et al., 1979).

16.8.2 Constructed treatment wetlands

Natural wetlands usually improve the quality of water passing through them, acting effectively as ecosystem filters. In comparison, most constructed wetlands are artificially created wetlands used to treat water pollution in its variety of forms. Therefore, they fall into the category of constructed treatment wetlands. Treatment wetlands are solar-powered ecosystems. Solar radiation varies diurnally, as well as on an annual basis (Kadlec, 1999).

Constructed wetlands have the purpose to remove bacteria, enteric viruses, SS, BOD, nitrogen (predominantly as ammonia and nitrate), metals and phosphorus (Pinney et al., 2000). Two general types of constructed wetlands are usually commissioned in practice: surface-flow (i.e. horizontal-flow) and sub-surface-flow (e.g., vertical-flow). Surface-flow constructed wetlands most closely mimic natural environments, and are usually more suitable for wetland species because of permanent standing water. In sub-surface-flow wetlands, water passes laterally through a porous medium (usually sand and gravel) with a limited number of macrophyte species. These systems have often no standing water.

Constructed treatment wetlands can be built at, above or below the existing land surface, if an external water source is supplied (e.g., wastewater). The grading of a particular wetland in relation to the appropriate elevation is important for the optimal use of the wetland area in terms of water distribution. Soil type and groundwater level must also be considered, if long-term water shortage is to be avoided. Liners can prevent excessive desiccation, particularly where soils have a high permeability (e.g., sand and gravel), or where there is limited or periodic flow.

Rooting substrate is also an important consideration for the most vigorous growth of macrophytes. A loamy or sandy topsoil layer between 0.2 and 0.3 m in depth is ideal for most wetland macrophyte species in a surface-flow wetland. A sub-surface-flow wetland will require coarser material such as gravel and/or coarse sand (Kadlec and Knight, 1996).

Furthermore, the use of flue-gas-desulphurisation by-products from coal-fired electric power plants in wetland liner material was researched, previously (Ahn et al., 2001). These by-products are usually sent to landfill sites. This is now recognised as an increasingly unsuitable and impractical waste disposal method. Although the study was short (two years), no detrimental impact on macrophyte biomass production was reported. Moreover, flue-gas-desulphurisation material may be a good liner and substrate for phosphorus retention in constructed wetlands.

The following conclusions with implications for treatment wetland design were made by Pinney et al. (2000):

- High levels of dissolved organic carbon may enter water supplies where soil aquifer treatment is used for groundwater recharge, as the influent for this method is likely to come from long hydraulic retention time wetlands.

Consequently, there is a greater potential for the formation of disinfection by-products.

- A shorter hydraulic retention time will result in less dissolved organic carbon leaching from plant material compared to a longer hydraulic retention time in a wetland.
- Dissolved organic carbon leaching is likely to be most significant in wetlands designed for the removal of ammonia, which requires a long hydraulic retention time.

16.8.3 Constructed wetlands for stormwater treatment

Most constructed wetlands in the USA and Europe are soil or gravel based horizontal-flow systems planted with *T. latifolia* and/or *P. australis*. They are used to treat urban runoff, domestic and industrial wastewater (Cooper et al., 1996; Kadlec and Knight, 1996; Scholz, 2003; Scholz et al., 2005), and have also been applied for passive treatment of mine wastewater drainage (Mays and Edwards, 2001; Mungur et al., 1997).

Storm runoff from urban areas has been recognised as a major contributor to pollution of the corresponding receiving urban watercourses. The principal pollutants in urban runoff are BOD, SS, heavy metals, de-icing salts, hydrocarbons and faecal coliforms (Scholz and Martin, 1998a; Scholz, 2004b).

Although various conventional methods have been applied to treat stormwater, most technologies are not cost-effective or too complex. Constructed wetlands integrated into a best management practice concept are a sustainable means of treating stormwater and prove to be more economical (e.g., construction and maintenance) and energy efficient than traditional centralised treatment systems (Kadlec et al., 2000; Scholz et al., 2005). Furthermore, wetlands enhance bio-diversity and are less susceptible to variations of loading rates (Cooper et al., 1996; Scholz and Trepel, 2004).

Contrary to standard domestic wastewater treatment technologies, stormwater (e.g., gully pot liquor and effluent) treatment systems have to be robust to highly variable flow rates and water quality variations. The stormwater quality depends on the load of pollutants present on the road, and the corresponding dilution by each storm event (Scholz, 2003; Scholz and Trepel, 2004).

In contrast to standard horizontal-flow constructed treatment wetlands, vertical-flow wetlands are flat, intermittently flooded and drained, allowing air to refill the soil pores within the bed (Cooper et al., 1996; Gervin and Brix, 2001; Green et al., 1998). While it has been recognized that vertical-flow constructed wetlands have usually higher removal efficiencies with respect to organic pollutants and nutrients in comparison to horizontal-flow wetlands, denitrification is less efficient in vertical-flow systems (Luederits et al., 2001). When the wetland is dry, oxygen (as part of the air) can enter the top layer of debris and sand. The following incoming flow of runoff will absorb the gas and transport it to the anaerobic bottom of the wetland.

Furthermore, aquatic plants such as macrophytes transport oxygen to the rhizosphere. However, this natural process of oxygen enrichment is not as effective as the previously explained engineering method (Kadlec and Knight, 1996; Karathanasis et al., 2003).

Heavy metals within stormwater are associated with fuel additives, car body corrosion, and tire and brake wear. Common metal pollutants from cars include copper, nickel, lead, zinc, chromium and cadmium. Freshwater quality standards are most likely to be exceeded by copper (Cooper et al., 1996; Scholz et al., 2002; Tchobanoglous et al., 2003).

Metals occur in soluble, colloidal or particulate forms. Heavy metals are most bioavailable when they are soluble, either in ionic or weakly complexed form (Cooper et al., 1996; Cheng et al., 2002; Wood and Shelley, 1999).

There have been many studies on the specific filter media within constructed wetlands to treat heavy metals economically; e.g., limestone, lignite, activated carbon (Scholz and Martin, 1998a), peat and leaves. Metal bioavailability and reduction are controlled by chemical processes including acid volatile sulphide formation and organic carbon binding and sorption in reduced sediments of constructed wetlands (Kadlec, 2002; Obarska-Pempkowiak and Klimkowska, 1999; Wood and Shelley, 1999). It follows that metals usually accumulate in the top layer (fine aggregates, sediment and litter) of vertical-flow and near the inlet of horizontal-flow constructed treatment wetlands (Cheng et al., 2002; Scholz and Xu, 2002; Vymazal and Krasa, 2003).

Physical and chemical properties of the wetland soil and aggregates affecting metal mobilisation include particle size distribution (texture), redox potential, pH, organic matter, salinity, and the presence of inorganic matter such as sulphides and carbonates (Backstrom et al., 2004). The cation exchange capacity of maturing wetland soils and sediments tend to increase as texture becomes finer because more negatively charged binding sites are available. Organic matter has a relatively high proportion of negatively charged binding sites. Salinity and pH can influence the effectiveness of the cation exchange capacity of soils and sediments, because the negatively charged binding sites will be occupied by a high number of sodium or hydrogen cations (Knight et al., 1999).

Sulphides and carbonates may combine with metals to form relatively insoluble compounds. Particularly the formation of metal sulphide compounds may provide long-term heavy metal removal, because these sulphides will remain permanently in the wetland sediments as long as they are not re-oxidised (Cooper et al., 1996; Kadlec and Knight, 1996).

Chapter 17

Rotating biological contactors

17.1 Introduction

Rotating biological contactors (RBC) were first installed in West Germany in the 1960s and later introduced to the UK and USA. Approximately 70% of RBC systems are removing carbonaceous BOD only, 25% are used for combined carbonaceous BOD removal and nitrification with the remainder being used for nitrification of secondary effluents. Rotating biological contactors consist of a series of closely spaced circular disks of polystyrene or polyvinyl chloride. The disks are submerged in wastewater and rotated slowly through it. Constructed wetlands can be used to polish the effluent of RBC.

17.2 Principle of operation

In operation, biomass (slime or film) grows on to the surface of the disks, and eventually covers the entire wetted surface. The rotation of the disks alternately contacts the biomass with organic matter in the wastewater and then with the atmosphere for the adsorption of oxygen.

The disk rotation affects oxygen transfer and maintains the biomass in an aerobic condition. The rotation is also the mechanism for removing excess solids from the disks by the shearing forces it creates as the disk passes through the wastewater. The sloughed or sheared biomass is then maintained in suspension, so that it can be removed from the unit to the sedimentation tanks. Some RBC are designed to allow the sloughed solids to settle to the bottom of the RBC tank and breakdown, the sludge being periodically removed.

17.3 Design and loading criteria

Although a few theoretical loading performance models have been developed, the usefulness of these models is still questioned by design engineers. As a consequence, the design of RBCs is largely based on the use of simple design parameters. These design parameters are derived from the experience of the operation of full-scale plants.

When RBC systems were originally introduced, the process design was based on hydraulic loading expressed in $m^3/m^2/d$ to achieve the required removal efficiency. Over the past 20 years, the design approach has shifted, first to the use of total BOD per unit of surface area (kg total $BOD/m^2/d$), and most recently to soluble BOD per unit of surface area per day (soluble $BOD/m^2/d$), or in the case of design for nitrification to kg $NH_3/m^2/d$.

Poor performance has been observed where systems are overloaded resulting in low DO concentrations in the units, H_2S odours and poor removal rates (particularly in the first stage). Typical design loading rates are shown in Table 17.1.

The design parameters are for mixtures of domestic wastewater with minor amounts of industrial (non-toxic) wastewater. Selection of a loading value within the noted range is made based on effluent requirements, temperature range, degree of uncertainty as to waste load and expected competency of the operating staff. The total media area is normally sized based on the mean annual design year conditions (unless data on significant loading variations throughout the year are available). After the total surface area required has been calculated, the design must be checked to avoid exceeding the oxygen transfer capacity of the first stage unit.

Table 17.1 Typical design data for rotating biological contactors at a wastewater temperature of 13°C

Item	Secondary	Combined	Separate nitrification
Hydraulic loading ($m^3/m^2/d$)	0.08–0.16	0.003–0.08	0.04–0.1
Organic loading (kg soluble $BOD_5/m^2/d$)	0.0037–0.01	0.0025–0.0075	0.005–0.015
Organic loading (kg total $BOD_5/m^2/d$)	0.01–0.0175	0.0075–0.015	–
Maximum load on the first stage (kg soluble $BOD_5/m^2/d$)	0.02–0.03	0.02–0.03	–
Maximum load on the first stage (kg total $BOD_5/m^2/d$)	0.04–0.06	0.04–0.06	–
NH_3 loading (kg $NH_3/m^2/d$)	–	0.00075–0.0015	0.001–0.004
Hydraulic retention time θ (d)	0.7–1.5	1.5–4.0	1.2–2.9
Effluent BOD_5 (mg/l)	15–30	7–15	7–15
Effluent NH_3 (mg/l)	–	<2	1–2

Note: BOD_5 = Five-day @ 20°C N-allyl thiourea biochemical oxygen demand; NH_3 = Nitrate-nitrogen

17.4 Principle elements

The principle elements of a RBC are as follows: shaft, media, drive system, tankage, enclosure and settling tank. A shaft is used to support and rotate the disks.

In practice, the shaft should not exceed 8.2 m, and have a maximum of 7.6 m occupied by the media. The minimum length occupied by the media is usually about 1.5 m. The major problem with shafts in RBC is shear failure due to the overload from excessive media growths.

Many different media have been used in RBC. The first plants used timber disks, though these have fallen out of fashion. Nowadays, the disk media are commonly manufactured from high density polyethylene, PVC or polystyrene, depending upon the manufacturer and the specification by the client.

Most RBC are driven by a direct mechanical drive. Air drive units are also available. They comprise a deep plastic cup attached to the perimeter of the media disk, and is driven by compressed air. Variable speed control is essential on all drive units.

The capacity of the tankage (i.e. space occupied by wastewater in the RBC) has been optimised at 0.0049 m^3/m^2 of media. This results in a hydraulic retention time of 1.44 h at a hydraulic loading rate of 0.08 $m^3/m^2/d$. A typical sidewater depth of 1.5 m is usually sufficient to give a 40% disk submergence.

Usually, RBC are covered with a fibreglass or reinforced plastic shell over each separate drive shaft assembly. This gives protection to the biofilm and plant in adverse weather conditions.

Settling tanks are not essential to the proper operation of a RBC. If they are provided, they should be designed as humus tanks.

17.5 Operational problems

Many of the early RBC units had operational problems. These were mainly shaft failures, media breakage, bearing failures and odour problems. Shaft failures have been the most serious equipment problem, because of the loss of a process unit from service and the damage to a portion of the media. The causes of shaft breakage may be attributed to inadequate structural design, metal fatigue and excessive biomass accumulation on the media.

Media breakage has been caused by exposure to heat, organic solvents, ultra-violet radiation or inadequate design of the media support systems. Bearing failures have been attributed to inadequate lubrication and poor maintenance. Odour problems are frequently caused by excessive organic loading rates, particularly in the first stage.

Modifications to the equipment have been made to mitigate against many of these problems and to simplify maintenance. When designing or specifying a RBC unit, it is essential to consult the manufacturer's literature.

The continuous mixing action is important not only to ensure that an adequate food supply gets to the micro-organisms, but also to support a maximum oxygen concentration gradient to enhance mass transfer, and to help disperse the metabolic waste products from within the flocs.

As the settled wastewater enters the aeration tank, it displaces the mixed liquor (i.e. the mixture of wastewater and microbial mass) into a sedimentation tank. This is the second stage, where the flocculated biomass settles rapidly out of suspension to form a sludge, and where the clarified effluent, which is virtually free from solids, is discharged as the final effluent.

In the conventional AS process, between 0.5 and 0.8 kg dry weight of sludge is produced for every kg of BOD_5 removed. The sludge is rather like a weak slurry containing between 0.5 and 2.0% dry solids, and so can be easily pumped. As the solids content increases, the viscosity rapidly becomes greater, although under normal operating conditions AS is difficult to concentrate to >4% dry solids by gravity alone.

Most of the AS is returned to the aeration tank to act as an inoculum of micro-organisms, ensuring that there is an adequate microbial population to fully oxidise the wastewater during its retention period within the aeration tank. The excess sludge (8–10% of that produced daily) requires treatment before disposal.

The most important function in the AS process is the flocculent nature of the microbial biomass. Not only do the flocs have to be efficient in the adsorption of and the subsequent absorption of the organic matter in the wastewater, but they also have to be rapidly and effectively separated from the treated effluent within the sedimentation tank. Any change in the operation of the reactor will lead to changes in the nature of the flocs, which can adversely affect the overall process in a number of ways; most notably, poor settlement (bulking) can result in turbid effluents and the loss of microbial biomass.

Although some variants of the AS process are used to treat sewage, which has only been screened and degritted, the majority of AS processes use settled sewage as the feedstock. The sludge produced from the process should not be confused with primary sludge (contains coarse organic and inorganic solids), as it is composed entirely of microbial biomass and adsorbed particulate matter.

The main components of all AS systems are as follows:

- Reactor: This can be a tank, lagoon or ditch. The main design criteria for a reactor is that its content can be adequately mixed and aerated. The reactor is also known as the aeration tank or basin.
- Activated sludge (AS): This is the microbial biomass within the reactor, which comprises bacteria and other microfauna and flora. The sludge is a flocculant suspension of these organisms, and is often referred to as the mixed liquor.

The normal concentration is estimated as MLSS (usually between 2000 and 5000 mg/l).

- Aeration and mixing system: Aeration and mixing of the AS and the incoming wastewater are essential. While these processes can be undertaken independently in separate tanks, they are normally combined with each other using a single system. Either surface aeration or diffused air is used.
- Sedimentation tank: Final settlement (or clarification) of the AS displaced from the aeration tank by incoming wastewater is required. This separates the microbial biomass from the treated effluent.
- Returned sludge: Some of the settled AS in the sedimentation tank is recycled back to the reactor to maintain the microbial population at a required concentration to ensure continuation of treatment.

Ideally, the AS process should be operated as close to a food-limited condition as possible to encourage endogenous respiration. This is where each micro-organism is utilising its own cellular content, thus reducing the quantity of biomass produced. During the endogenous respiration phase, the respiration rate will fall to a minimum value, which is sufficient for cell maintenance only. However, under normal operating conditions, the growth of the microbial population and the accumulation of non-biodegradable solids result in an increase of the amount of AS produced.

The two principle removal mechanisms in the AS process are as follows:

- Assimilation: Utilisation of waste to create new biomass. Colloidal and soluble BOD is transformed into biomass, which is then settled out.
- Mineralisation: Oxidation (degradation) of waste to inert end products, which are either vented to the atmosphere or are left in solution in the effluent.

Activated sludge plants can be operated to favour either of these two processes. In plants dominated by assimilation, there is rapid removal of the BOD with a correspondingly high production of sludge, which means that sludge treatment costs are high. Where plants are operated as to favour mineralization, then these require long aeration times. Thus the operating costs increase due to increased air or oxygen requirements. This is offset by the reduced production of sludge, and correspondingly low sludge treatment costs.

Whichever method is favoured, the wastewater itself must contain adequate nutrients for biological growth. This nutrient requirement is usually expressed in terms of the carbon, nitrogen and phosphorus ratio (Eq. (18.2.1)).

$$BOD_5 : N : P = 100 : 5 : 1$$

(BOD$_5$ being used as a measure of the carbon content) (18.2.1)

18.3 Comparison between the activated sludge process, percolating filtration and wetland system

The relative merits and demerits of the AS process, percolating filtration (Chapter 15) and wetland system (Chapter 16) parameters are summarised in Table 18.1.

Table 18.1 Comparison of the activated sludge (AS) process, percolating filtration and wetland system parameters

	AS process	Percolating filtration	Wetland system
Capital cost	Low	High	Very low
Area of land	Low; advantageous where land availability is restricted or expensive	Large; ten times more area required	Large; at least ten times more area required
Operating cost	High	Low	Very low
Influence of weather	Works well in wet weather, slightly worse in dry weather, less affected by low winter temperatures	Works well in summer but possible 'ponding' in winter	Independent of weather, if the wetland is sufficiently large
Technical control	High; the microbial activity can be closely controlled; requires skilled and continuous operation	Little possible except process modifications; does not require continuous or skilled operation	Little possible; does not require continuous or skilled operation
Nature of wastewater	Sensitive to toxic shocks and changes in loading; trade wastewaters can lead to bulking problems	Strong wastewaters acceptable; able to withstand changes in loading and toxic discharges	Strong wastewaters acceptable; able to withstand changes in loading and toxic discharges
Hydrostatic head	Small; low pumping requirement; suitable for a site where the available hydraulic head is limited	Large; site must provide natural hydraulic head, otherwise pumping is required	Small; suitable for a site where the available hydraulic head is limited
Nuisance	Low odour and no fly problems; noise may be a problem in both urban and rural areas	Moderate odour; severe fly problem in summer; quiet	No unacceptable nuisance due to an 'ecological equilibrium'
Final effluent quality	Poor nitrification but low in suspended solids, except when separation problems occur	Highly nitrified; relatively high suspended solids	High-quality effluent unless the system is overloaded

Table 18.1—cont'd

	AS process	Percolating filtration	Wetland system
Secondary sludge	Large volume; high water content; difficult to dewater; less stabilised	Small volume; less water; highly stabilised	Virtually no sludge production (except for detritus) within the system's lifetime
Energy requirement	High; required for aeration, mixing and maintaining sludge flocs in suspension and for recycling sludge	Low; natural ventilation; gravitational flow	Virtually no energy required
Synthetic detergents	Possible foaming; especially with diffusers	Little or no foam	No problem
Robustness	Not very robust; high degree of maintenance on motors; not possible to operate without power supply	Very sturdy; low degree of maintenance; possible to operate without power	Very sturdy; low degree of maintenance; possible to operate without power

18.4 Activated sludge process types

18.4.1 Conventional complete mix activated sludge process

Since its introduction, the AS process has undergone many variations and adaptations. In some cases, these adaptations have resulted from basic research into the principles of the process, but more usually they have arisen as empirical solutions to particular problems in plant operation.

Concerning the conventional AS process, also called 'complete mix', the effluent from primary sedimentation tanks is aerated in the aeration tank. This aeration usually lasts 6–12 h at 3 DWF. The MLSS are kept fairly high in order to increase the efficiency of the biological reactions. Typical MLSS are between 3 and 3.5×10^3 mg/l. The organic loading rate for these units is between 3 and 3.5 kg $BOD_5/m^3/d$. The air supplied to the aeration unit is of the order of 6 m^3/m^3 sewage. This type of plant does not produce a nitrified effluent.

The effluent quality of the treated sewage after 2 h of sedimentation is as follows: BOD_5 between 10 and 15 mg/l, and SS of 20 mg/l. The aeration in the aeration tank is achieved at a reasonably uniform rate over the length of the tank. This leads to a possible oxygen shortage at the inlet. Reactions may therefore be oxygen-limited. Hence, the process efficiency may be reduced. The conventional system provides good buffering against shock loads and toxic conditions; and is thus particularly suitable for the treatment of industrial effluents.

18.4.2 Series or plug flow system

In this arrangement, the waste to be treated and the return AS are introduced at one end of a channel, in which a number of aerators are located, and treated mixed liquor is withdrawn at the other end. With mechanical aerators, there may be as many as ten aerators in a line, but it is usual to have more lines with three or four aerators per line. Compared to a single line, this provides a higher initial degree of aeration to the untreated waste. Domestic sewage is commonly treated by this method as long as there are no inhibitory substances present. One advantage of this arrangement is that a progressively reducing substrate level produces a sludge with improved settling characteristics.

18.4.3 Tapered aeration

This is a modification of the plug flow system. The process seeks to adjust the rate of oxygen supply throughout the length of the tank to the rate of oxygen demand. Hence, a higher proportion of the total air supply is introduced at the inlet end, and the rate of supply is tapered (reduced) towards the tank outlet. The rate at the outlet is normally set to a value approaching that of endogenous respiration. This system is particularly suited to the treatment of strong readily biodegradable wastes.

18.4.4 Step feed AS process

This is another variation of the plug flow process, which attempts to equalise the oxygen supply and demand. In this case, instead of varying the rate of oxygen supply along the aeration tank, partial equalisation of the demand is achieved by adding the waste to the aeration tank at various points along the length of the tank. There is no apparent advantage for this system concerning the treatment of domestic sewage.

18.4.5 High-rate activated sludge process

This is sometimes called 'modified aeration' process. This type of plant operates at very much higher food to micro-organism ratios than the conventional process:

- Conventional: 0.2–0.4 kg BOD_5/kg MLVSS/d; and
- High rate: 1.5–5.0 kg BOD_5 kg/MLVSS/d.

This high loading rate is achieved by a much reduced hydraulic retention time (about 2 h) and a much lower MLSS (about 1000 mg/l). Air supply is 3 m^3/m^3 with BOD_5 reductions between 60 and 70% being possible. The process results in more of the incoming organic matter being synthesised to sludge organisms.

Therefore, the total oxygen requirements are somewhat less than in conventional AS, but the rate of oxygen demand is higher per unit of MLVSS.

18.4.6 Extended aeration

This variant is characterised by a very low food-to-micro-organism ratio, low net sludge yield, higher hydraulic retention times and higher MLSS. This makes the process suitable for small isolated communities.

The disadvantages of aeration systems for small communities are as follows:

- The effect of high hydraulic loading causing loss of AS solids;
- Widely fluctuating BOD loads; and
- The uneconomic demands of conventional aeration systems for regular attention (e.g., surplus AS management).

In the extended aeration plant, primary sedimentation can be omitted, so that only sludge from the wastage line is to be disposed off. It follows that this process is ideal for rural locations. The extended aeration system provides an aeration tank capacity of >24 h DWF, which is a large buffer for high instantaneous loadings. Nevertheless, the final settlement tanks should be designed for the treatment of peak loads.

The loading rate applied to these systems is between 0.24 and 0.32 kg $BOD_5/m^3/d$. The hydraulic retention time is between 24 and 36 h. This should produce an effluent of typically 40 mg/l BOD_5 and 50 mg/l SS.

One advantage of the extended aeration system is that the aeration process is taken well into the endogenous respiration stage. Consequently, the sludge yield is minimised. The sludge age is ≥ 10 d, which means that the effluent is at least partially nitrified.

18.4.7 Contact stabilisation

Contact stabilisation tries to exploit the rapid reduction in BOD of the untreated waste by biosorption and bioflocculation with the return AS. In this system, the untreated waste and return AS are aerated in a contact stage with a retention between 0.5 and 3 h at DWF. The sludge, separated by settlement, is passed to a stabilisation tank where it is re-aerated for a period usually <6 h based on the AS return flow to complete the oxidation of the adsorbed BOD.

The degree of purification obtained in the contact zone depends on the nature of the waste, the contact time and the MLSS concentration. As the proportion of BOD in true solution increases, the contact period must be increased until the process may become conventional. An effluent with a BOD of <20 mg/l and SS of <30 mg/l can be achieved for domestic sewage with this treatment process.

they have a lower aeration efficiency, but the advantages of lower cost, less maintenance and the absence of stringent air purity requirements offset the slightly lower efficiency.

In the static tube aerator, air is introduced at the bottom of a circular tube, which can vary in height between 0.5 and 1.25 m. Internally, the tubes are fitted with alternately placed deflection plates to increase the contact of the air with the wastewater. Mixing is accomplished, because the tube aerator acts as an airlift pump.

Other types of diffuser are available on the market. Jet aeration, for example, combines liquid pumping with air diffusion. The pumping system recirculates liquid in the aeration basin, and ejects it with compressed air through a nozzle assembly. This system is useful for deep tanks (6–7 m). Aspirating aeration consists of a motor-driven aspirator pump. The pump draws air in through a hollow tube, and injects it underwater, where both high velocity and propeller action create turbulence, which diffuses the air bubbles. U-tube aeration is used in the deep shaft process. Typical transfer efficiencies for these devices range between 9 and 40%.

18.5.2 Mechanical aerators

This class of aerators is divided into two groups, aerators with a vertical axis and aerators with a horizontal axis. Both groups can be further classified as surface or submerged aerators. Surface aerators with a vertical axis are designed to induce either updraft or downdraft flows through pumping action.

High speed aerators are used in ponds and lagoons. The water level varies so that a fixed support is impractical. Therefore, nearly always floating devices are used in practice.

Low speed aerators are used in the AS process, usually on a fixed platform. They are designed to throw wastewater into the atmosphere as small droplets. Sometimes, they are fitted with a draft tube to improve efficiency.

Submerged devices with a vertical axis rely on violent agitation of the surface, and subsequent air entrainment to achieve enhanced oxygen transfer. These are not popular devices, though they have been used in pure oxygen systems.

Mechanical aerators with a horizontal axis can also be sub-classified into surface and submerged devices. The surface aerator types are based on the so-called Kessener brush aerators, a device normally used in oxidation ditches (Chapter 18.4.8). The brush type aerator has a horizontal axis with bristles mounted just above the water surface. The bristles are then submerged, and the cylinder rotated rapidly, spraying wastewater across the tank, promoting circulation and entraining air.

Submerged aerators with a horizontal axis are similar to the surface type aerators, except that they use disks or paddles attached to the rotating shaft to agitate the fluid. The disk aerators have been used previously in oxidation ditches.

The disks are submerged to approximately between 0.125 and 0.375 of their diameters. Recesses on the disks introduce entrapped air beneath the surface as the disks turn. Typical efficiency figures for mechanical aerators are in the range between 8 and 15%.

18.5.3 Process design

18.5.3.1 Kinetics of biological growth
An empirically developed relationship between biological growth and substrate utilisation, which is commonly used in biological systems stabilising organic and inorganic waste, is expressed in Eq. (18.5.2).

$$\text{Net growth rate of micro-organisms} = \text{birthrate} - \text{deathrate.} \qquad (18.5.2)$$

The food-to-microorganism ratio is known as the process loading factor, the substrate removal rate and the specific utilisation. The fractional growth, the micro-organism growth rate and the food-to-microorganism growth rate are directly related with each other. However, the following assumptions have to be made:

- All the necessary nutrients for growth are present;
- Temperature and pH are regulated to achieve the optimum growth rate; and
- The equations apply only to that portion of the waste, which is soluble or biodegradable.

18.5.3.2 Application of kinetics
The four principle types of reactors (i.e. tank systems) for biological wastewater treatment are classified according to their hydraulic characteristics as batch, plug, complete mix (continuous but stirred flow) and arbitrary flow.

In plug flow, fluid particles pass through the tank, and are discharged in the same sequence in which they enter. The particles retain their identity, and remain in the tank for a time equal to the theoretical detention time. Complete mixing occurs when particles entering the tank are immediately dispersed throughout the tank. The particles leave the tank in proportion to their statistical population. The batch reactor is characterised by the fact that flow neither enters, nor leaves the reactor on a continuous basis. Arbitrary flow represents any degree between partial mixing and plug flow, and is difficult to describe mathematically. Hence, ideal plug flow or complete mix flow models are usually assumed.

18.5.3.3 Complete mix reactor (no recycle)
The reactor is completely mixed and there are no organisms in the influent. The hydraulic retention time and the mean cell residence time are the same. The mass balance for the micro-organisms in the reactor is as follows: the 'rate of change of

organism concentration in the reactor' is equal to the net 'rate of organism growth in the reactor' substracted by the 'rate of organism outflow from the reactor'. The micro-organism concentration is a function of the mean cell residence time.

The complete mix (no recycle) model can be used to simulate conventional anaerobic treatment systems (usually sludge digestion), and some modified AS processes, oxidation ponds and lagoons, provided an allowance is made for settling where appropriate.

The treatment efficiency of a process with a high cell residence time can be limited by the depletion of nutrients, oxygen transfer difficulties and problems with the mixing of the large micro-organism mass resulting from extended cell residence times.

18.5.3.4 Complete mix cellular reactor (recycle)

It is assumed that the reactor content is completely mixed, and that there are no micro-organisms in the waste effluent. Since such a system involves a settling unit, further simplifying assumptions must be made, namely:

- Waste stabilisation occurs only in the reactor; and
- The volume used to derive the mean cell residence time only includes the reactor volume.

For this system, the mass balance will be as follows: the 'rate of change of organism concentration in the reactor' is equal to the 'net rate of organism growth in the reactor' substracted by the 'rate of organism outflow from the reactor'.

There is neither a need to determine the amount of biological solids in the system, nor the amount of food utilised. A specified percentage of the cell mass in the system must be wasted each day to control the growth rate of micro-organisms, and hence the degree of waste stabilisation.

Unlike the complete mix (recycle) scheme, in most biological treatment processes, cell wastage occurs from the sludge recycle line (surplus AS). Both the mixed liquor and return sludge micro-organism concentration should be known, if cells are wasted from the sludge recycle line.

18.5.3.5 Plug flow (cellular recycle)

In a plug flow model, all particles entering the reactor will remain there for an equal amount of time. Thus, simplifying assumptions can be made. For example, the influent micro-organism concentration is approximately equal to the effluent micro-organism concentration. Furthermore, the excess micro-organisms are wasted from the reactor outflow and not from the recycle line.

Theoretically, the plug flow recycle system is more efficient at waste stabilisation than is the complete mix recycle system. In practice, however, longitudinal dispersion prevents a true plug flow regime. The plug flow system is also more susceptible to 'shock' loads (see below). Therefore, the aeration tank is normally divided into a series of complete mix reactors (i.e. combination of plug flow with complete mix) to give better treatment together with improved resistance to 'shock' loads.

18.6 Summary of activated sludge processes

18.6.1 Loading criteria

The factors to be considered in the design of AS systems are as follows:

- Loading criteria;
- Selection of the reactor type;
- Sludge production;
- Oxygen requirements and transfer;
- Nutrient requirements;
- Environmental requirements;
- Solid–liquid separation; and
- Effluent characteristics.

The following parameters are used to control and design the AS process:

(1) Food-to-microorganism ratio; and
(2) Mean cell residence time.

Generally, the food-to-microorganism ratio is between 0.2 and 0.5. Values for the mean cell residence time are between 7 and 15 d, and result in a stable and high quality effluent. The corresponding sludge should be easy to dewater.

Empirical relationships based on detention time and volume have also been used. Detention times may be based on the waste flow disregarding recirculation (sewage detention period) or on the mixed liquor flow inducing recirculation (mixed liquor detention period).

This empirical design method is based on the consideration of the maximum organic loading in terms of kg BOD/m^3/d, together with the specification of the detention time. The method ignores the concentration of the mixed liquor, the food-to-microorganism ratio, and the mean cell residence time. However, it has the merit of requiring a minimum aeration tank volume that should be adequate for satisfactory treatment, assuming proper selection of other design.

18.6.2 Reactor types

Initial construction, operation and maintenance costs will affect the reactor selection process in practice. Operational factors, which must be considered, include the following:

- The reaction kinetics governing the treatment process;
- Oxygen transfer requirements;
- The nature of the wastewater to be treated; and
- Local environmental conditions.

For first-order substrate removal kinetics, the total volume required for a series of complete mix reactors (i.e. four or more) is considerably less than that required for a single complete mix reactor. Indeed the volume differential becomes more pronounced as the removal efficiency increases, and also as the order of substrate removal kinetics increases. In practice, however, neither a plug flow nor a complete mix reactor functions as assumed in theory.

With conventional plug flow aeration systems, insufficient oxygen is supplied to meet the oxygen requirements of the input and of the reactor. Modifications to improve the oxygen transfer route can be summarised as follows:

- Tapered aeration, where the air supply is matched to the oxygen demand;
- Step aeration, where the incoming waste and the return solids are distributed along the length of the reactor; and
- A complete mix AS process, where the aeration matches the oxygen demand.

In a complete mix reactor, the incoming waste is uniformly dispersed. As a result, it can withstand shock loads (i.e. discharges of a sudden and/or toxic nature) better than a plug flow reactor. Thus, the complete mix reactor is the preferred system for industrial waste treatment.

Theoretically, the most important environmental variable is temperature. However, the conventional AS process is not significantly influenced by temperature in practice. It should be noted that AS processes are generally less sensitive to low temperatures than trickling filters (Chapter 15.2).

The quantity of sludge produced is vital to the design of the sludge handling and disposal facilities. The volumetric sludge production depends on the volume of the reactor, and the efficiency of the final settling tank.

18.6.3 Oxygen demand

The oxygen requirement is calculated from the BOD (Chapter 4.2) of the waste, and from the quantity of cells wasted from the system per day. Assuming that all the BOD is converted to end products, the total oxygen demand can be computed by converting BOD_5 to BOD_{ult} and subtracting the BOD_{ult} of the cells wasted from the system. Using the molecular weight relationships for typical waste ($C_5H_7NO_2$) and the associated oxygen requirement, it follows that the oxygen demand (kg/d) can be estimated by the food utilised per day reduced by 1.42 times the organisms wasted per day.

If the oxygen transfer efficiency of the aeration system is known, the air requirement can be determined. The minimum dissolved oxygen concentration throughout the reactor should be between 1 and 2 mg/l. Generally, an aeration system should be capable of delivering 150% of the normal air requirement.

18.6.4 Nutrient requirements

Adequate nutrients must be available, if the plant is to function properly. The principal nutrients for all biological treatment systems including constructed wetlands are nitrogen and phosphorus. Assuming a cell composition of $C_5H_7NO_2$, then 12.4% nitrogen by weight of the mass of organisms produced each day is required. The corresponding phosphorus demand is about 0.2 of the nitrogen demand. In general, wastewater contains all nutrients required for cell growth. However, industrial wastes may require additional nutrients.

Chapter 19

Iron and manganese removal

19.1 Introduction

Selecting an appropriate process sequence for the removal of iron and manganese from water supplies, for example, is made complex by the variety of reactions that can occur with these elements. The choice depends on the presence or absence of manganese, carbon dioxide, organic acids, turbidity, DO, hardness and bicarbonates. It also depends on the binding of Fe and Mn to organic matter (e.g., peat) and also on the pH. In general, pilot plant studies are essential to optimize the processes.

19.2 Problems with iron and manganese

Iron and manganese have to be removed from some waters for various reasons:

- They cause corrosion and pipe blockages (directly by precipitation or indirectly by creating favourable growth conditions);
- They affect the appearance of the water;
- They impart a metallic taste to the water;
- They cause laundering difficulties; and
- They pose a potential pollution problem to the receiving watercourse.

Particular problems may be experienced with groundwater having a high Fe and Mn content. These precipitate on contact with air and turn dark brown or black, creating obvious problems.

19.3 Basic removal processes

Iron alone in groundwater, which contains little or no organic material, can be removed by simple aeration, followed by detention, sedimentation and filtration.

However, it is usually carried out in towers containing slats or trays, which are either seeded with catalytic material such as pyrolusite (MnO_2) or which are permitted to accumulate deposits of Fe_2O_3. The latter deposits also serve to catalyse the oxidation of iron. The detention and sedimentation basin common to such systems is unlikely to accumulate much in the way of solids. Its purpose is primarily to provide sufficient time for the oxidation reaction to reach completion, and for iron to come out of solution. The precipitated iron is removed by filtration. A number of manufacturers provide packaged iron removal systems, which incorporate an aeration tower, a small detention tank and a filter in a single unit. This type of system can be completely satisfactory for small flows with the characteristics described above.

If both iron and manganese are present or if the water contains organic material such as humic or fulvic acid, aeration is sufficiently rapid only if it is catalysed by pyrolusite or by accumulation of oxidation products (Fe_2O_3 and MnO_2) on a porous bed such as coke. Simple aeration will not provide oxidation and subsequent precipitation within a reasonable time, although elevation of the pH will increase the precipitation rate substantially. Manganese is much more slowly oxidised than iron. In fact, the rate is negligible at pH levels <9. Organic material interferes with the removal of iron and manganese by peptisation; i.e. by complexing with the metal without complete neutralisation of the molecular charge, by forming soluble complexes with both the reduced and oxidised metals, by reduction of the oxidised metals, or by a combination of these processes.

The application of strong oxidising agents such as chlorine, ozone, chlorine dioxide or potassium permanganate can serve to oxidise iron and manganese more rapidly, and can also oxidise or destroy the organic materials present. Such oxidising agents have the potential of forming trihalomethanes from humic and fulvic acid, and thus must be applied with care. Permanganate in this application will function adequately at neutral pH. The other oxidants require pH levels of about 9.5 for adequate manganese removal. Iron is oxidised satisfactorily at neutral pH, but its oxidation is enhanced under more alkaline conditions; e.g., at a pH between 8.0 and 8.5, and followed by the removal of the oxidised metals via rapid filtration (Chapter 13).

19.4 Advanced removal processes

Catalytic action may be required for the oxidation of manganese. This occurs in beds of either pyrolusite ore or zeolite regenerated with potassium permanganate. In some cases, potassium permanganate is fed continuously into the water before it enters the filter. A bed of coke coated with deposits of ferric hydroxide can also act as a catalyst.

Lime softening increases the pH. Preliminary aeration removes carbon dioxide, and results in the conversion of iron to its ferric state, and subsequently its precipitation as ferric hydroxide.

Ion exchange using either zeolites or other resins can remove ferrous and manganous ions, if they are associated with bicarbonate. These conditions occur most frequently with oxygen-deficient groundwaters. Care must be taken to avoid aeration or oxidation of such waters, as this would produce Fe^{3+} and Mn^{4+} ions, which will precipitate and clog the ion exchange bed. This type of process is practicable, when used as an incidental to softening only, and when combined Fe and Mn is less than 0.3 times the hardness.

Sequestration of Fe and Mn may be satisfactory in some circumstances. This involves the use of complex molecules, which encase (or combine with) the Fe and Mn ions, so that they, although still in solution, can no longer take part in other reactions. The usual sequestering agents are polyphosphates or organic compounds such as sodium hexametaphosphate (e.g., marketed as Calgon). Subsequent heating of the water may destroy the effectiveness of this treatment, so it must be used with care.

Most wetland systems can also be used to remove iron and manganese. The main processes are detention, sedimentation and predominantly filtration.

Chapter 20

Water softening

20.1 Introduction

Softening is practised when the hardness of the outflow from a wetland system, for example, is not acceptable; the degree of softening depends upon the use to which the water will be put. For domestic water supply, for example, hardness creates nuisance scums, wastes soap when used for laundry or ablution, and deposits scale in hot water services. Softening is seldomly applied to municipal water supplies, if the hardness is ≤ 150 mg/l as $CaCO_3$; and a reasonable target after treatment is a hardness of approximately 80 mg/l as $CaCO_3$.

For general treatment of municipal supplies, a process of precipitating the calcium and magnesium hardness by adding lime ($Ca(OH)_2$) and/or soda (caustic soda (NaOH) or soda ash (Na_2CO_3)) is usually adequate.

Hardness is caused by calcium and magnesium ions in the water. The presence of bicarbonate ions affects the dose of chemicals required to precipitate the calcium and magnesium, and therefore hardness is classified in two ways, namely:

- As calcium or magnesium hardness; and
- As carbonate or non-carbonate hardness.

Carbonate hardness is caused by calcium or magnesium ions associated with bicarbonate ions, and non-carbonate hardness is caused by calcium or magnesium ions associated with chloride, sulphate or other ions.

20.2 Chemistry of water softening

The chemical reactions involved in water softening are as follows (Eqs. (20.2.1) and (20.2.2)):

$$CaO + H_2O \rightarrow Ca(OH)_2 \tag{20.2.1}$$

$$CO_2 + Ca(OH)_2 \rightarrow \underline{CaCO_3} + H_2O \qquad (20.2.2)$$

For carbonate hardness with calcium (Eq. (20.2.3)):

$$Ca(HCO_3)_2 + Ca(OH)_2 \rightarrow \underline{2CaCO_3} + 2H_2O \qquad (20.2.3)$$

For non-carbonate hardness with calcium (Eq. (20.2.4)):

$$CaSO_4 + Na_2CO_3 \rightarrow \underline{CaCO_3} + Na_2SO_4 \qquad (20.2.4)$$

For carbonate hardness with magnesium (Eq. (20.2.5)):

$$Mg(HCO_3)_2 + 2Ca(OH)_2 \rightarrow \underline{Mg(OH)_2} + \underline{2CaCO_3} + 2H_2O \qquad (20.2.5)$$

For non-carbonate hardness with magnesium (Eq. (20.2.6)):

$$MgSO_4 + Ca(OH)_2 + Na_2CO_3 \rightarrow \underline{Mg(OH)_2} + \underline{CaCO_3} + Na_2SO_4 \qquad (20.2.6)$$

For excess lime removal (Eq. (20.2.7)):

$$Ca(OH)_2 + Na_2CO_3 \rightarrow \underline{CaCO_3} + 2NaOH \qquad (20.2.7)$$

For carbon dioxide removal (Eq. (20.2.8)):

$$CO_2 + Ca(OH)_2 \rightarrow \underline{CaCO_3} + H_2O \qquad (20.2.8)$$

Equations (20.2.1)–(20.2.8) show important reactions, which can take place in the softening process. Inspection of these chemical equations shows that the required dose may be estimated as follows:

Equivalents of lime = equivalents of carbonate hardness
+ equivalents of magnesium hardness
+ equivalents of carbon dioxide
+ excess lime to raise the pH

Equivalents of soda = equivalents of non-carbonate hardness
+ equivalents of excess of lime

20.3 Lime-soda softening

The quantities of soda ash (Na_2CO_3) and lime ($Ca(OH)_2$) are estimated from a chemical analysis of the water, and are subsequently added to the influent stream. Sometimes, a small amount of iron salts or alum is added as a flocculation aid. The dosed water passes to a sludge blanket clarifier (or a similar treatment unit), where the chemicals precipitate out, and are subsequently removed. Lime is a cheap form of alkali, which will raise the pH to between 9.5 and 10.0, so that bicarbonates will be converted to carbonates and the calcium precipitates as calcium carbonate.

Further raising of the pH will cause the magnesium hardness to precipitate as magnesium hydroxide. This is achieved by adding an excess of lime over that indicated as needed by the chemical equations, and of sufficient soda to precipitate the calcium leaving sodium hydroxide in the water. The amount of excess lime added is often between 40 and 65 mg/l as $CaCO_3$ depending on the desired efficiency of magnesium removal. As the pH is increased, any CO_2 present will be converted to carbonate, and will be available for the precipitation of calcium. Water with a high CO_2 concentration can be aerated before lime treatment to reduce the chemical requirements by eliminating some of the CO_2.

Lime-soda softening is used for all forms of calcium hardness. By adding soda-ash, non-carbonate hardness is converted to $CaCO_3$, which will then precipitate. As much soda ash is added as there is non-carbonate hardness associated with calcium.

Excess lime-soda softening is used for all forms of magnesium hardness. The chemistry is complex, and the required dose rate is best determined experimentally.

20.4 Lime softening

Lime softening is used for calcium hardness of the carbonate form. The addition of lime (equivalent to the amount of bicarbonate present) will form the insoluble calcium carbonate. The solubility of $CaCO_3$ at normal temperatures is about 20 mg/l. Considering the limited contact time available in a standard plant, a residual of about 40 mg/l can be measured. Thus, the water must be stabilised prior to filtration. The water leaving has a fairly high pH (9.5–11). If it were applied to the filter, calcium carbonate would precipitate in the filter, and cement the sand grains together.

The usual method of reducing the pH, and bringing the water to a stable condition in which it will not precipitate calcium carbonate, is to add CO_2, usually by burning gas, oil or coke. The products of combustion are bubbled through the water in a recarbonation tank. If too little CO_2 dissolves, the tendency to precipitate $CaCO_3$ is increased. In contrast, if too much CO_2 dissolves, the water can become corrosive.

In cases where the target hardness and raw water quality are suitable, split treatment can eliminate the need for recarbonation and results in the saving of chemicals. However, extra tanks increase capital costs. In split treatment, one stream is dosed with lime and softened by the excess lime process; some of the calcium and most of the magnesium in that stream are removed. The flows are then recombined and dosed with soda ash, and, if necessary, passed to a second clarifier. The caustic alkalinity of the water, which has been treated with excess lime, reacts with the bicarbonates in the untreated stream to form carbonates. These carbonates react with the remaining calcium to form $CaCO_3$, most of which will precipitate. Split treatment removes practically all the magnesium from the first stream, but practically none from the second. Recarbonation can be expressed with Eq. (20.4.9):

$$CaCO_3 + CO_2 + H_2O \rightarrow Ca(HCO_3)_2 \qquad\qquad (20.4.9)$$

20.5 Excess lime softening

Excess lime softening is used for the removal of magnesium carbonate hardness. The above methods are not effective for the removal of magnesium, since magnesium carbonate is soluble (Eq. (20.5.10)).

$$Mg(HCO_3)_2 + Ca(OH)_2 \rightarrow \underline{CaCO_3} + MgCO_3 + 2H_2O \qquad (20.5.10)$$

However, at about pH 11, the following process occurs (Eq. (20.5.11)):

$$MgCO_3 + Ca(OH)_2 \rightarrow \underline{Mg(OH)_2} + \underline{CaCO_3} \qquad\qquad (20.5.11)$$

The practical solubility of $Mg(OH)_2$ is approximately 10 mg/l. For excess lime softening, it is necessary to add $Ca(OH)_2$ equivalent to HCO_3^-, $Ca(OH)_2$ equivalent to Mg^{2+} and 35 mg/l excess $Ca(OH)_2$ to raise the pH to 11.

Carbonation is necessary to remove the excess lime, and to reduce the pH after treatment. The high pH also produces disinfection as a by-product, and so lower chlorine doses may be necessary after softening in a water treatment works, for example.

20.6 Lime recovery

All forms of precipitation softening produce considerable volumes of sludge. Lime recovery is possible by calcining $CaCO_3$ sludge and subsequently slaking CaO with water as shown in Eqs. (20.6.12) and (20.6.13). In this way, more lime than is required in the plant is produced, and the surplus may be sold. This solves the sludge disposal problem at the same time.

$$CaCO_3 \rightarrow CaO + CO_2 \text{ (CO}_2 \text{ can be used for recarbonation)} \qquad (20.6.12)$$

$$CaO + H_2O \rightarrow Ca(OH)_2 \qquad (20.6.13)$$

Chapter 21

Water microbiology

21.1 Statistics for applied microbiology

The quantitative and qualitative assessment of wetland system microbial dynamics is important when estimating the treatment performance and assessing public health issues. There are three common ways to estimate the average of a population; averages can be described as mean, median and mode values. The mean is equivalent to the average. The median is the middle observation in a set of observations, which have been ranked in magnitude. The (crude) mode is the class in a frequency distribution, which contains more observations than any other.

In a symmetrical distribution, mean, median and mode are equivalent. In contrast, the quantitative relationship in a skewed distribution is as follows: mean > median > mode. Skewed distributions may require data transformation in order to perform advanced statistical tests on the observations. The estimate of the standard deviation for a population is defined in Eq. (21.1.1).

$$s = \sqrt{\frac{\sum(x - \bar{x})^2}{n - 1}} \qquad (21.1.1)$$

where,

s = population;
x = the value of an observation;
$\bar{x}$ = the sample mean; and
n = the number of observations.

The standard deviation is the square root of the variance, which describes the variability of a sample. A logarithmic transformation (x is replaced by $\log_{10} x$), in order to convert skewed into symmetric data, is recommended, if the variance of a sample of count data is larger than the mean. In contrast, the square root transformation (x is replaced by $\sqrt{x}$) is appropriate when the variance of a sample of count data is close to the mean. Other data transformation methods including the arcsine transformation are less popular in praxis.

The number of species present in a sample is a very simple measure of the diversity. However, there are a number of other indices, which take into account how evenly the total number of individuals in a sample is apportioned between each species. The fundamental diversity index can be expressed as a ratio of the number of species and the number of individuals. Furthermore, the Shannon–Weaver Diversity Index, a popular (often applied) but controversial (difficult to interpret) population diversity index is defined in Eq. (21.1.2).

$$H = -\sum_{i}^{s} p_i \ln p_i \qquad (21.1.2)$$

where,

H = Shannon–Weaver Diversity Index;
S = total number of species; and
p_i = proportion of a particular species in a sample.

For practical purposes, a diversity index can be regarded as a number on an ordinal scale. It is sensible to use diversity indices only to compare 'like with like'. Comparisons between diversity indices are valid only if the samples from which the indices are calculated have approximately the same number of observations.

21.2 Protozoa

21.2.1 Trophic structure

Protozoa have key roles in both the autotrophic and heterotrophic food chains of wetland systems (Scholz and Martin, 1998a,b):

Autotrophic food chain (fuelled principally by sunlight)

Producers
>   Photosynthesis: algae, floating plants, attached forms and rooted plants.
>   Chemosynthesis: bacteria.

Primary consumers
>   Grazers (herbivores): protozoa, rotifers, crustaceans and other
>   herbivores.

Secondary consumers
>   Carnivorous: small fish.

Tertiary consumers
>   Carnivorous: large fish.

Heterotrophic food chain (fuelled by organic matter)
Particle producers including bacteria.
Detritus including dead organisms, excretions and faeces.
Decomposers (including bacteria and fungi)
Primary consumers
    Protozoa.
Secondary consumers
    Rotifers.
Tertiary consumers
    Fish.

21.2.2 Kingdom Protista

Living things can be divided into five kingdoms: Monera, Protista, Plantae, Fungi and Animalia. Protista can be subdivided as follows:

Phylum I: Sarcomastigophora
Flagella and/or pseudopodia present; spores not produced.

Subphylum 1: Mastigophora
Flagella typically present, division by longitudinal binary fission.

Class A: Phytomastigophorea
Plant-like flagellates typically with chloroplasts; if missing, relationship to pigmented forms clearly evident (e.g., *Euglena* and *Peranema*)

Class B: Zoomastigophorea
Animal-like flagellates; chloroplasts absent; one to many flagella; amoeboid forms; with or without flagella; many parasitic forms (e.g., *Bodo* and *Giardia*).

Subphylum 2: Opalinata
Binary fission takes place between rows of cilia, which cover the entire body in oblique rows; two or many monomorphic nuclei; all parasitic; none associated with polluted waters.

Subphylum 3: Sarcodina
Pseudopodia typical; flagella restricted to developmental stages when present.

Superclass A: Rhizopoda
Locomotion by pseudopodia or by protoplasmic flow without production
of discrete pseudopodia. It includes the naked (*Acanthamoeba, Amoeba,
Entamoeba* and *Naegleria*) and testate amoebae (*Arcella* and *Euglypha*),
the foraminifera and the slime-mould amoebae.

Superclass B: Actinopoda
Spherical; typically planktonic; axopodia with delicate internal micro-
tubular skeleton; some naked, others with tests of chitin, silica or
strontium sulphate (e.g., *Actinophyrys*).

Phylum II: Apicomplexa
Apical complex visible with an electron microscope; all species are parasitic.

Class A: Perkinsea
No sexual reproduction; incomplete cone; none in polluted waters.

Class B: Sporozoea
Sexual and asexual reproduction typical; oocysts generally containing
invective sporozoites, which result from sporogony. Locomotion of
mature organisms by gliding. It includes the gregarines and coccidia
(e.g., *Cryptosporidium*).

Phylum III: Microspora
Unicellular spores each with an imperforated wall, containing one uninucleate or
dinucleate sporoplasm; always with polar tube and cap; obligatory intracellular
parasites in nearly all major animal groups.

Phylum IV: Ciliophora
Cilia or compound ciliary organelles present in at least one stage of the life cycle;
two types of nucleus present; sexual and asexual reproduction; plane of division
transverse across ciliary rows.

Class A: Kinetofragminophorea
Oral cilia slightly distinct from body ciliature; cytostome often apical or
mid-ventral on surface of body; body ciliation commonly uniform (e.g.,
Acineta, Amphileptus, Chilodonella, Colpoda and *Litonotus*).

Class B: Oligophymenophorea
Oral apparatus, at least partially in buccal cavity, generally well
defined, although absent in one group; oral ciliation clearly distinct
from body cilia; cytostome usually ventral at or near anterior end at

bottom of buccal cavity (e.g., *Carchesium, Colpidium, Glaucoma, Paramecium, Uronema* and *Vorticella*).

Class C: Polyhymenophorea
Dominated by well-developed, conspicuous adoral zone of numerous buccal or peristomial ciliary organelles; cytostome at bottom of buccal cavity (e.g., *Aspidisca, Euplotes* and *Metopus*).

21.3 Biological effects of organic pollutants

21.3.1 Sewage fungus

Sewage fungus is present in organically polluted waters commonly found at the beginning of a wetland system. Most samples of sewage fungus contain protozoa (mainly flagellates and ciliates). Colourless micro-flagellates including *Bodo caudatis* are the most common flagellated protozoa, although a number of species of *Euglena*, and its colourless relative *Peranema trichophorum*, are sometimes also observed.

The most important ciliates are *Colpidium colpoda* and *Chilodonella cucullulus*, which swim or glide between and along the filamentous material. Some of the ciliates including *Chilodonella* are known to feed on filamentous organisms. Most of the others feed on suspended bacterial populations. Some ciliates (e.g., *Litonotus* and *Trachelophyllum*) are carnivorous and prey upon other ciliates. Furthermore, diatoms are frequently present in slimes (*Navicula, Fragilaria* and *Synedra*). Organisms most frequently found in sewage fungus growth are listed below.

Bacteria
Zoogloeal bacteria, *Sphaerotilus natans, Flavobacterium* spp., *Beggiatoa alba, Flexibacterium* sp. and *Sirillum* sp.

Blue-green 'algae'
Oscillatoria spp. and *Phormidium* spp.

Fungi
Geotrichum candidum, Leptomitus lacteus and *Fusarium aquaeductum.*

Algae
Stigeoclonium tenue and *Ulothrix* spp.

Diatoms
Navicula spp., *Fragilaria* spp. and *Synedra* spp.

Protozoa
Ciliated protozoa
Hemiophrys fusidens, Litonotus fasciola, Trachelophyllum pusillum, Chlodonella cucullulus, Chilodonella uncinata, Colpidium campylum, Colpidium colpoda, Glaucoma scintillans, Cinetochilum margaritaceum, Paramecium trichium, Carchesium polypinium, Vorticella convallaria, Aspidisca cicada, Aspidisca lynceus and *Tachysoma pellionella.*
Other protozoa
Colourless micro-flagellates, thecate amoebae and naked amoebae.

21.3.2 Saprobic system

A simple saprobic system can be applied for most treatment wetland systems to indicate the different zones of polluted waters:

Polysaprobic zone
High concentration of biodegradable organic compounds; low DO concentration; high concentrations of ammonia and hydrogen sulphide; high BOD; low micro-organism diversity; bacteria, 'sewage fungus' community, turbificid worms and chironomid larvae are present.

Mesosaprobic zones
α-mesosaprobic zone: less organic material present; preponderance of amino acids rather than proteins; ammonia concentration has been reduced but nitrates have appeared; blanket weed (*Cladophora*) increases but other polysaprobic organisms decrease; slow recovery of dissolved oxygen.

β-mesosaprobic zone: recovery of dissolved oxygen and increase of population diversity; declining blanket weed.

Oligosaprobic zone
Self-purification complete; organic matter has been degraded and mineralisation completed; great diversity of species but reduced numbers of bacteria.

Flagellates and bacterivorous protozoa are more frequent in the polysaprobic to mesosaprobic zones while carnivorous ciliates are more frequent in the oligo-saprobic zones. The following sequence of saprobic zones and corresponding protozoa can be identified in a river receiving sewage:

Polysaprobic zone (p): *Bodo, Colpidium, Glaucoma* and *Paramecium.*
α-mesosaprobic zone (m or α): *Paramecium, Carchesium* and *Vorticella.*
β-mesosaprobic zone (m or β): *Vorticella, Chilodonella, Spirostomum, Didnium* and *Coleps.*

Oligosaprobic zone (o): *Didinium, Coleps* and *Litonotus*.

The five major objections related to the limitations of the use of saprobic systems are as follows:

(1) Highly trained personnel are required to identify microorganisms;
(2) Incomplete knowledge about microorganisms;
(3) Samples from aquatic habitats have to be statistically representative;
(4) A saprobic system generalises pollution and watercourses; and
(5) The original saprobic system is limited to pollution caused by sewage discharge to rivers.

21.4 Eutrophication and water treatment

The input water of water treatment works is often polluted with large numbers of algae, in particular diatoms, during spring. Table 21.1 gives an example of different levels of pollution by algae for lake and reservoir water.

The effects of eutrophication on the receiving ecosystem can be summarised as follows:

• Species diversity often decreases and the dominant biota subsequently changes;
• Plant and animal biomass increases;
• Turbidity increases;
• Rate of sedimentation increases, shortening the life span of the wetland; and
• Anoxic conditions may develop.

Table 21.1 Characteristic algal associations of oligotrophic and eutrophic lakes and reservoirs

State	Algal group	Examples
Oligotrophic	Picoplankton (often small cyanobacteria)	*Synechococcus*
	Desmid plankton	*Staurdesmus* and *Staurastrum*
	Chrysophycean plankton	*Dinobryon*
Mesotrophic	Diatom plankton	*Cyclotella* and *Tabellaria*
	Dinoflagellate plankton	*Peridinum* and *Ceratium*
	Chlorococcal plankton	*Oocystis* and *Eudorina*
Eutrophic	Diatom plankton	*Asterionella, Fragilaria, Stephanodiscus* and *Melosira*
	Dinoflagellate plankton	*Peridinium* and *Glenodinium*
	Chlorococcal plankton	*Scenedesmus*
	Cyanobacterial plankton	*Alphanizomenon, Anabaena* and *Microcystis*

The problems to man associated with the effects of eutrophication on the ecosystem are summarised below:

- The treatment of water may be difficult, and the supply may have an unacceptable taste and odour;
- The water may be injurious to health;
- The amenity value of the water may decrease;
- Increased vegetation may impede water flow and navigation; and
- Commercially important species such as salmonids and coregonids (whitefish) may disappear.

Typical trophic classification system data are summarised in (Table 21.2). Most constructed treatment wetlands are eutrophic or hypertrophic.

In trickling filters, suspended solids including plankton are partly consumed directly by macro-invertebrates, and suspended bacteria by protozoa and rotifers. Annelids, collembola and the majority of fly larvae are important in preventing fixed film filters being blocked during warm temperatures.

21.5 Protozoology of treatment processes

Protozoa are plentiful in constructed treatment wetlands, AS and other polluted mixed liquors (up to about 50 000 cells per ml). Protozoa species and their abundances can be related to water and wastewater treatment processes, and their corresponding water quality.

The spatial distribution of protozoa in a filter depends on the level of saprobity. It is possible to monitor the filter effluent quality by interpreting protozoa species and levels in various parts of the filter. The temporal succession of protozoa groups within polluted waters that become gradually cleaner is as follows: flagellates → free-swimming ciliates → crawling ciliates → attached ciliates (Scholz and Martin, 1998a,b).

Table 21.2 Trophic classification system variables

Trophic category	TP	Mean Chl	Maximum Chl	Mean Secchi	Minimum Secchi
Ultra-oligotrophic	<4.0	<1.0	<2.5	>12.0	>6.0
Oligotrophic	<10.0	<2.5	<8.0	>6.0	>3.0
Mesotrophic	10–35	2.5–8	8–25	6–3	3–1.5
Eutrophic	35–100	8–25	25–75	3–1.5	1.5–0.7
Hypertrophic	>100	>25	>75	<1.5	<0.7

TP = mean annual total phosphorus concentration (μg/l); mean Chl = mean annual chlorophyll a concentration in surface waters (μg/l); maximum Chl = peak annual chlorophyll a concentration in surface waters (μg/l); mean Secchi = mean annual Secchi depth transparency (m); minimum Secchi = minimum annual Secchi depth transparency (m).

Protozoa may be used as indicator organisms to predict treatment plant efficiencies in terms of BOD effluent concentrations (Scholz and Martin, 1998a,b). Under protozoa-free conditions, plants produce effluent of low quality with respect to BOD, permanganate value, organic nitrogen, suspended solids, viable bacteria, COD and optical density.

21.6 Odour and toxins of natural origin

There are four main groups of algae, which are able to cause odour problems: blue-green algae, diatoms, green algae and yellow-brown algae.

(1) The important genera of blue-green algae (*Cyanophyceae*) are *Anabaena*, *Anacystis*, *Aphanizomenon*, *Gomphosphaeria* and *Oscillatoria*. These generally give rise to grassy odours intensifying into piggy, almost septic, odours as the cells disintegrate. Some algae can also produce toxic substances.
(2) The most important genera of diatoms (*Bacillariophyceae*) are *Asterionella*, *Fragillaria*, *Melosira*, *Tabellaria* and *Synedra* (Scholz and Martin, 1998b). They secrete oils, which impart a fishy or aromatic odour. When moderate densities of *Asterionella* and *Tabellaria* are present, then geranium odours are produced.
(3) Green algae (*Chlorophyceae*) can impart a fishy or grassy odour. Important genera include *Volvox* and *Staurastrum*.
(4) Yellow-brown algae (*Chrysophyceae*) can produce pungent odours. *Uroglenopsis* can produce a very strong fishy odour, whereas *Synura* produces a cucumber odour at low, a fishy odour at moderate and a piggy odour at high numbers.

It is the blue-green algae, which are responsible for toxins in drinking waters. Despite their name, these algae are actually a group of bacteria, which are capable of photosynthesis. It follows that blue-green algae are also called cyanobacteria.

There are no water quality standards set for algal toxins. However, water companies monitor toxic algal blooms regularly. Most algal blooms can be prevented by adding copper sulphate to the untreated water. Activated carbon treatment helps to oxidise algal toxins (Scholz and Martin, 1998a).

21.7 Public health aspects

21.7.1 Typical diseases related to waters

Wetland systems might harbour organisms that are potentially pathogenic. Protozoa are known to feed upon pathogenic bacteria, including those which cause diseases (e.g., diphtheria, cholera, typhus and streptococcal infections), and faecal bacteria (e.g., *Escherichia coli*). Water-borne infectious diseases are

- *Nematoda*, roundworms or eelworms. They are the most common larger micro-organisms in water supply systems. Large numbers of bacteria and an elevated BOD concentration can be correlated with the presence of nematodes.
- Smaller *Crustacea* (e.g., *Cyclops* and *Chydorus*). They are common in water mains.
- Terrestrial animals (e.g., slugs and earthworms). They die rapidly in waters.

The presence of some invertebrates does correlate well with high counts of certain bacteria in potable water distribution networks; e.g., *Nematoda* correlate well with *Pseudomonas cepacia*. Furthermore, fly larva are associated with *Aeromonas hydrophila*, *Pseudomonas cepacia* and *Pseudomonas fluorescens*, and copepods with *Flavobacterium meningosepticum*, *Moraxella* spp., *Pasteurella* spp., *Pseudomonas diminuta* and *Pseudomonas paucimobilis*.

Finally, amphipods are associated with many bacteria genera and species in water distribution systems; they include *Acinetobacter* spp., *Achromobacter xylooxidaans*, *Bacillus* spp., *Chromobacter violaceum*, *Moraxella* spp., *Pasteurella* spp., *Pseudomonas fluorescens*, *Pseudomonas maltophilia*, *Pseudomonas paucimobilis*, *Pseudomonas vesicularis*, *Serratia* spp. and *Staphyloccocus* spp.

21.7.3 Monitoring and prevention of water-borne diseases

Relevant criteria for potential indicator micro-organisms of faecal contamination are as follows:

- The micro-organisms should be a member of the normal intestinal flora of healthy people;
- They should be exclusively intestinal in habitat and therefore exclusively faecal in origin, if found outside the intestine;
- Ideally, they should only be found in humans;
- They should be present when faecal pathogens are present, and only when faecal pathogens are expected to be present;
- They should be present in greater numbers than the pathogen that they are intended to indicate;
- An indicator micro-organism should be unable to grow outside the intestine with a death rate slightly less than the pathogenic organism;
- They should be resistant to natural environmental conditions, and to water and wastewater treatment processes in a manner equal to or greater than the pathogens of interest;

- They should be easy to isolate, identify and enumerate; and
- They should preferably be non-pathogenic.

Traditionally, the following maximum admissible conditions apply for most finished waters:

- Zero total coliforms per 100 ml sample, where sufficient numbers of samples are examined, then a 95% consistent result is acceptable;
- Zero faecal coliforms (*E. coli*) per 100 ml;
- Zero faecal streptococci per 100 ml sample;
- Less than one sulphite reducing clostridia per 20 ml sample; and
- There should be no significant increase in the total bacteria colony counts above background levels, although a guide value of <10 per ml at 22°C and <100 per ml at 37°C has been set.

Furthermore, guidelines require that there is no significant increase from background levels of heterotrophic bacteria in either tap- or bottled waters. Heterotrophic plate counts represent the aerobic and facultative anaerobic bacteria that derive their carbon and energy from organic compounds. This group includes Gram-negative bacteria belonging to the following genera: *Pseudomonas*, *Aeromonas*, *Klebsiella*, *Flavobacterium*, *Enterobacter*, *Citobacter*, *Serratia*, *Actinetobacter*, *Proteus*, *Alcaligenes* and *Moraxella*.

Changes in heterotrophic bacteria counts indicate a variation in water quality, and the potential for pathogenic survival and regrowth. They are also applied to monitor disinfection efficiencies. Changes in the pattern of colony counts are more significant than single numerical counts for the interpretation of bacterial activity within a treatment process.

Heterotrophic plate counts are normally done by using the spread plate count method using Yeast Extract Agar (incubated at 22°C for 72 h and 37°C for 24 h). Results are expressed as colony forming units. Counts at 37°C are particularly useful as they can provide a rapid indication for possible contamination of water supplies.

Both physical and biological potable water filtration systems are effective in cyst removal. When rapid physical filtration systems are used, it is essential that filters are adequately maintained and operated, preferably with a pre-treatment of chemical coagulation and sedimentation procedures.

Rapid sand filters remove 99.99% of *Entamoeba histolytica* cysts. However, these figures cannot be expected for poor conditions of coagulation. In contrast, diatomaceous earth filters can essentially remove all cysts. As with the control of giardiasis, the usual recommended procedure for the safe control of amoebiasis is to use some form of wetland filtration, with prior chemical coagulation, followed by chlorination.

Pathogenic bacteria have been associated with protozoa. Under specific environmental conditions, some protozoa can be related indirectly to certain diseases. Certain pathogenic bacteria can live in the cytoplasm of free-living, and non-pathogenic protozoa. For example, the ciliates *Tetrahymena* and *Cyclidium* can act as hosts to *Legionellae*.

Chapter 22

Disinfection

22.1 Destroying pathogens and requirements of a disinfectant

Disinfection of the outflow water from wetland systems might be required for particular applications. Disinfection is carried out to destroy the microbiological agents, which cause disease. It differs from sterilisation, which implies the destruction of all micro-organisms. Water supplies are usually disinfected. The widespread adoption of disinfection is a major factor in reducing water-borne diseases, and this has been interpreted as the major single factor in increasing average human life expectancy.

Chlorination, usually performed as the final treatment process, is the most common means of disinfecting predominantly drinking water. Other processes, natural and artificial, aid in the destruction or removal of pathogens.

Most pathogens are accustomed to the temperatures and conditions in the bodies of humans and animals; they do not survive outside such environments. Nonetheless, significant numbers can survive in potable water. Some pathogens, particularly certain viruses and those organisms that form cysts, can survive for surprisingly long periods even under the most adverse conditions. Because such organisms also tend to be resistant to chlorine doses normally used in water treatment, chlorination alone cannot always ensure safe drinking water. Storing water for extended periods of time in open tanks or reservoirs prior to treatment can accomplish some destruction of pathogens through sedimentation and natural death of the organisms. Significant pathogen removal occurs in the water and wastewater treatment processes.

A disinfectant must be able to destroy particular pathogens at the concentrations likely to occur, and it should be effective in the normal range of environmental conditions. Disinfectants, which require extremes of temperature or pH, or which are effective only for low-turbidity waters, are unlikely to be suitable for large-scale use.

While the disinfectant should destroy pathogens, it must not be toxic to man or other higher animals (e.g., fish) in a receiving watercourse. Ideally, some residual disinfecting capacity should be provided for a water supply network to provide protection against re-infection while the water is in the distribution system. The residual, which passes to the consumer, should be neither unpalatable nor obnoxious.

A disinfectant should be safe and easy to handle, both during storage and addition to the process water. The other major consideration is that of cost, which is particularly important for municipal plants where large volumes are commonly disinfected continuously.

22.2 Traditional methods of disinfection

Although chlorination is the most common technique for disinfection, other methods are available and may be useful in some situations. The three general types of disinfection commonly available are as follows:

- Heat treatment;
- Radiation treatment; and
- Chemical treatment.

Heat was probably the first method used for disinfection, and it is still an excellent method in emergencies for small quantities of water. Boiling water for 15–20 min (depending on altitude) destroys pathogens. Pasteurisation, which is essentially heating to a temperature $\geq 60°C$, is effective in destroying many pathogens as well. Moreover, autoclaving, where the temperature is $\geq 100°C$ (typically $121°C$) at increased pressure, can also be used for the disinfection of small-scale wastewater treatment plants.

Heat is uneconomical for large-scale use, primarily because of the high cost of energy required. However, it is acceptable and commonly used in emergency situations. The water contains no residual, and care must be taken to prevent recontamination before use.

Concerning radiation treatment, sunlight (high-intensity ultraviolet radiation) is a natural means of disinfection. It works by breaking down long-chain organic molecules, and subsequently kills pathogens.

Since ultraviolet (UV) light is readily absorbed by water and 'filtered' by turbidity, the normal penetration depth of radiation is less than 120 mm. This makes the process useful for small-scale disinfection only. No residual is provided, so care against re-infection must be exercised.

Ultraviolet disinfection of water is growing in its usage. In both the USA and Australia, UV is used to disinfect recycled water prior to its use for irrigation or landscaping purposes. The technology is now at a stage where the lamps are self-cleansing, and the power and number of tubes needed for a given log kill can be predicted.

22.3 Ozone

Ozone (O_3) is an allotrope of oxygen. It is formed when a high voltage arc passes through the air between two electrodes. It is also formed photochemically in the atmosphere, and is one of the constituents of smog. Ozone is a bluish and toxic gas with a pungent odour. It is considered hazardous to health at a concentration of 0.25 mg/l in air (by volume) and is extremely hazardous at levels of ≥ 1 mg/l. It is corrosive in nature. It follows that only resistant materials like stainless steel and polyvinyl chloride (PVC) can be used to handle it. Ozone is unstable, because it breaks down to give molecular oxygen. Its low solubility and instability requires that it is to be generated on site and introduced into the water as fine bubbles.

A retention time between 5 and 10 min is usually sufficient to destroy pathogens. Ozone is an effective method of destroying pathogens. It is a strong oxidant, and is effective for removing taste, odour, iron, manganese and colour constituents from waters. Its ability to completely oxidise organic materials minimises the organic content of water, and so inhibits the growth of low life forms in water. It is not known to produce dangerous or objectionable compounds, as chlorine does in yielding chlorophenols and trihalomethanes.

The only product from the addition of excess ozone is the breakdown product oxygen, and so nothing foreign is added. The breakdown of ozone does mean that there is no lasting residual disinfectant in the water. If a residual is required, it may be necessary to add chlorine or chlorine dioxide to act as a residual in the distribution system, while ozone is used as the primary disinfectant, and for its other effects such as taste and odour suppression.

Ozone acts through the destruction of all protoplasm. Disinfection by ozone depends mainly upon the concentration of the gas added. Once the concentration of ozone falls below a critical value for each pathogen, its effectiveness as a disinfectant is significantly reduced. This behaviour differs from that of chlorine.

Ozone is usually more expensive than chlorine; it requires a high energy input during production (approximately 10–15 times that of chlorine). It is only slightly soluble in water, and leaves no residual. The difficulties associated with the production of toxic or noxious compounds by reaction with chlorine have led to a renewed interest in ozone as a disinfectant for water and effluents.

22.4 Chlorine dioxide

Chlorine dioxide (ClO_2) is an unstable gas, which can be generated by the addition of acid or chlorine to sodium chlorine. In water, it reacts to produce two unstable acids; chlorous acid and chlorite acid, which can act as disinfectants. Chlorine dioxide is a highly reactive gas, and particular care is required to ensure its safe handling. Its instability requires it to be generated on site, and to be used immediately.

Although chlorine dioxide is relatively expensive as a disinfectant, its strong oxidising characteristics make it valuable for the control of taste, odour, iron, manganese and colour. Chlorine dioxide is not affected by ammonia compounds. It oxidises compounds to form tasteless products, and provides a persistent residual in the distribution system.

22.5 Chlorine as a disinfectant

Chlorine is available in gaseous, liquid and solid forms. Under normal conditions, chlorine is a green-yellow and corrosive gas with a density 2.5 times that of air. It can be liquefied under relatively small pressure (3.7 atm). The gas is very soluble in water, and is a potent disinfectant even at low concentrations. When dissolved in water, chlorine forms two acids by reaction with molecules of water, as shown in Eq. (22.5.1).

$$
\begin{array}{ccccc}
 & & & & \text{free available chlorine} \\
Cl_2 + H_2O & \longrightarrow & HCl & + & HOCl \\
\text{(pH = 9)} & & \uparrow\downarrow & \text{(pH = 5)} & \uparrow\downarrow \\
 & & H^+ Cl^- & & H^+ ClO^- \\
 & & & & \text{(free residuals)}
\end{array}
$$

$$(22.5.1)$$

Hydrochloric acid (HCl) is a common mineral acid and plays no significant role in disinfecting water. Hypochlorous acid (HOCl) is a disinfecting agent, and is referred to as free available chlorine. Both acids can dissociate to produce their respective anions and protons. For hypochlorous acid, this dissociation is important, since the hypochlorite ion is not a disinfectant. The hyperchlorous acid is dissociated by 50% at pH 7.5. To ensure effective disinfection, the pH must be acidic, so that at least 90% is in the undissociated and active disinfectant form (at pH 5, it is completely undissociated).

Chlorine rapidly penetrates microbial cells, and subsequently kills the corresponding micro-organisms through cell lysis. However, the effectiveness of chlorine is influenced greatly by the physical and chemical characteristics of the waste. The presence of SS or the clustering of micro-organisms may protect pathogens, and so reduce the disinfecting ability. Chlorine is a strong oxidising agent and any reducing agents such as nitrites, ferrous ions and hydrogen sulphite, rapidly react with it, reducing the concentration available to destroy pathogens.

Ammonia is present in both water and effluents. It reacts with chlorine in water to form sequentially monochloramine, dichloramine and trichloarmine. The reactions can be considered as successive replacements of hydrogen atoms in ammonia by chlorine (Eqs. (22.5.2) to (22.5.4)).

$$NH_3 + HOCl \rightarrow H_2O + NH_2Cl \qquad \text{(monochloramine)} \qquad (22.5.2)$$

$$NH_2Cl + HOCl \rightarrow H_2O + NHCl_2 \qquad \text{(dichloramine)} \qquad (22.5.3)$$

$$NHCl_2 + HOCl \rightarrow H_2O + NCl_3 \qquad \text{(trichloramine)} \qquad (22.5.4)$$

In addition, several other reactions may occur. The monochloramine and dichloramine may react further to give nitrogen gas (N_2) and/or nitrous oxide (N_2O). The proportions of any of these products depend upon the reaction conditions, notably concentrations, temperature and pH.

Chlorine reacts rapidly with any reducing agents in water. This reduces chlorine to chloride, which is not a disinfectant. The residual chlorine is low, and there will be no disinfecting properties for a small chlorine dose.

The addition of more chlorine leads to complete oxidation. Monochloramines and dichloramines are formed by reaction with ammonia. The chloramines are disinfected, and are detected as a chlorine residual. However, they are much less powerful disinfectants than free chlorine. If they are present in high concentrations or if a long reaction time is permitted, these chloramines will destroy pathogens. Monochloramine and dichloramine are referred to as combined available chlorine. The addition of further chlorine produces an approximately proportional increase of combined chlorine residual.

Even further addition of chlorine produces trichloramine, nitrogen, nitrous oxide and other products, which are not disinfectants. Therefore, the addition of further chlorine reduces the available chlorine, and hence the ability of the solution to destroy pathogens.

On further addition of chlorine, these reactions are complete, and the ammonia is completely oxidised. Any subsequent addition of chlorine will remain as free available chlorine (HOCl), which will act as a strong chlorine residual, if the pH is adequate. This point is referred to as the 'breakpoint chlorination', and is commonly carried out beyond the breakpoint to ensure a free chlorine residual. The chlorine dose required to reach the breakpoint is several times the ammonia concentration.

The chlorine–ammonia reaction (described above) should be regarded as a simplified representation. The free and combined residual concentrations depend upon the reaction rate and boundary conditions for any particular sample.

22.6 Kinetics of chlorination

The ability of a reagent to destroy pathogens is related to the concentration of the disinfectant, and the contact time between the disinfectant and pathogens. The distribution of micro-organisms is controlled by processes of diffusion.

The rate of kill depends upon the number of micro-organisms, which were originally present. If the micro-organisms all possess the same resistance, the kill follows an exponential pattern. A complete kill is not feasible. The efficiency of disinfection is reported in terms of the ratio of micro-organisms killed to the number of micro-organisms originally present; e.g., 99%, 99.9% or 99.99% kills, or sometimes also referred to as 2, 3 or 4 log kills.

The kill rate depends upon such factors as the penetration of the cell wall, the time to penetrate vital centres within the cell, and the distribution of the disinfectant and micro-organisms. Therefore, each micro-organism species is likely to have a different sensitivity to each disinfectant.

For disinfectants such as chlorine, the contact time is usually the most important single factor to ensure adequate pathogen destruction. A contact time of 0.5 h (or up to 1 h) with a residual afterwards of between 0.2 and 0.5 mg/l is the norm in practice.

22.7 Applications of chlorine

Chlorine can be used directly as a disinfectant, in which case the active reagent is hypochlorous acid. Alternatively, combined chlorine residuals can be formed, and they can also act as disinfectants.

Typical contact times for hypochlorous acid in waters of satisfactory purity (e.g., post filtration) are around 0.5 h. Chloramines are weaker disinfectants, and require contact times of the order of 1 h.

When chlorine is added to a particular water, the residual is a function of both the dose and the time elapsed. The necessary conditions for an adequate kill of pathogens are incorporated in the dosing requirements; e.g., the total free residual after x h shall be not less than y mg/l.

In addition to disinfection, chlorine has some further benefits for the water supply industry. It is a strong oxidant, capable of oxidising some materials in the water. Colour, taste and odour in water supplies are often of biological origin, and are caused by the presence of organic molecules. Iron and manganese salts may be present particularly in groundwater. They are usually in their reduced forms, and they precipitate out in the distribution system. Chlorine will oxidise some of these materials, reducing colour, taste and odour, and converting ferrous to ferric ions, which can be removed as a precipitate.

The presence of phenolic compounds in water supplies is a particular problem. Phenols (C_6H_5OH) react with chlorine to form chlorinated phenols, which have a very penetrating taste and odour. Conventional chlorination will produce chlorophenols, and it is therefore unsuitable, if phenols are present. The use of chloramines sometimes avoids this problem.

The other use of chlorine in water treatment is as a prechlorination stage prior to rapid sand filtration. This reduces the bacterial load on filters, and enables longer filter runs before backwashing is needed. It provides additional safety, in that a lower number of pathogens will be present for final disinfection, and growths within the treatment works are also controlled.

22.8 Technology of chlorine addition

Chlorine may be purchased as solid calcium hypochlorite, as liquid sodium hypochlorite or as liquid chlorine. Liquified chlorine, available in drums or cylinders, is used at large treatment plants. The rate of removal of chlorine from a drum or cylinder is a function of temperature. The vapour pressure of chlorine is temperature dependent and, as liquid chlorine vaporises, latent heat for vaporisation must be supplied. The temperature of a cylinder will be less than the average air temperature as the latent heat is absorbed. If chlorine is withdrawn too rapidly from a cylinder or drum, frosting or excessive sweating due to condensation occurs. In very damp climates, the rates of removal should be reduced.

Therefore, with a high air temperature, it is possible to withdraw chlorine relatively quickly, whereas for a low air temperature, slow withdrawal is needed. For example, it may be possible to use an entire cylinder in 4 d at an air temperature of 21°C, while 30 d would be needed at 2°C. At high temperatures, frothing of gas (while boiling) causes a carry over of impurities. Condensation of chlorine in gas lines should be avoided. A pressure-reducing valve attached near the cylinder or drum should keep the pressure in the pipes below the vapour pressure associated with the pipe temperature.

Chlorine is dissolved in a small part of the flow to be disinfected. This produces a strong solution of hypochlorous acid, which can then be added to the main stream to provide the required chlorine dose. The mechanism for dosing chlorine into the service water to produce strong hypochlorous acid requires careful handling and control.

The chlorinator includes a pressure-reducing valve, vacuum and pressure-relief valve, and an anti-suction valve. The high solubility of chlorine in water, combined with the hydraulic design of the injector, ensures that chlorine is drawn into the water under vacuum. This reduces the possibility of release of chlorine to the atmosphere, if the equipment develops a leak. The hypochlorous acid solution can be released into a bulk water stream by turbulent diffusion. For example, at a Reynolds number of ≥ 4000, the chlorine will be dispersed well within approximately ten standard pipe diameters. The chlorine can be dispersed at points of hydraulic discontinuity; e.g., near a submerged weir.

In practice, it is most convenient to control chlorination by adjustment of the chlorine concentration. Other possible variables, such as the contact time and form of chlorine residual, are fixed by the design of the plant, and the quality of the water. The control vacuum is varied manually or automatically on the basis of the measured flow rate and the determination of chlorine residual.

22.9 Advantages and disadvantages of chlorine

Chlorine is the most widely used chemical disinfectant. It has found widespread use not only for large and small water supply systems, but also for the treatment of swimming pool waters and effluent disinfection. Chlorine is a powerful disinfectant at low concentrations. Its ability to destroy pathogens without adversely affecting the future use of the water is an advantage. Analytical methods have been developed, which provide a simple, direct and accurate measurement of chlorine residuals. It is a dangerous chemical, which requires careful handling. However, the widespread and long-standing use of chlorine has resulted in the establishment of well-documented controls for safe operation. Moreover, it can be produced in large quantities relatively cheaply.

There are some disadvantages in the use of chlorine. Its reaction with phenolic compounds can make it unsuitable in some circumstances. Both free chlorine and chloramines have some adverse effect upon some higher animals (notably fish). Chlorination has been shown to produce trihalomethane compounds such as chloroform (CH_3Cl) by reaction with organic matter in the water. These compounds are known as carcinogens. Chlorine is not effective in removing certain tastes, odours and colours from water.

Chapter 23

Sludge treatment and disposal

23.1 Introduction

Wetland systems produce considerable amounts of sludge and sediment. It can be said that the main objective in wastewater treatment with treatment wetlands is to convert a water-based pollution problem to one that concerns solid waste management. The treatment and disposal of sewage solids (sludge) is a major component in the overall treatment costs, and should therefore be considered. The treatment and disposal of sludge has historically been an area that has attracted little attention. However, as costs continue to rise and the available sites for disposal decrease, sludge treatment has begun to become an area of interest for environmental engineers.

23.2 Characteristics of wastewater sludges

The main objective of all forms of wastewater treatment is the production of an effluent suitable for disposal into the environment or for some form of reuse. It follows that the impurities present in the wastewater must be either transformed into innocuous end products, or be effectively separated from the effluent stream. Impurities are either drawn off as a side stream to the main flow or converted into gaseous products. Treatment and disposal of side streams is an essential part of the overall treatment process, and they frequently contribute significantly to the total cost of the treatment.

In wastewater treatment works, the main side stream products (apart from screenings and grit) are the various forms of sludge from the underflows of sedimentation tanks. Treatment and disposal of these sludges is dependent on the volume and characteristics of the sludges.

The simplest classification of wastewater sludges is based on the process from which they are produced. Raw or primary sludge is drawn from the primary sedimentation tanks. It contains the readily settleable matter from the wastewater, has a high organic content (mainly faecal matter and food scraps) and is therefore highly putrescible. In its fresh state, raw sludge is grey in colour and has a heavy faecal odour. Both colour and odour intensify on prolonged storage under anoxic conditions, leading rapidly to the onset of putrefaction and very unpleasant odours. This is often evident in small works, when sludge is drawn from the sedimentation tank into an open pit for transfer to a digestion tank.

Humus sludge comprises the underflow from the humus tanks, which is usually located after a trickling filter. It consists of biological solids sloughed off or scoured from the surface of the filter media, and thus represents the net growth of biomass in the filters. The sloughing rate varies throughout the year, but is usually most intense during spring. Being largely organic in content, humus sludge exhibits problems similar to those of primary sludge under anoxic or anaerobic conditions.

Excess activated sludge (AS) is usually drawn off the return AS stream from the underflow of the secondary sedimentation tanks in the AS treatment process. It consists of light flocculent biological solids with a significant demand for oxygen, largely owing this demand to the respiration of the sludge micro-organisms. The mass of excess sludge produced each day represents the net growth in biomass, and is related to the loading rate of the process. Typically, AS has a much higher moisture content than other organic wastewater sludges, and therefore it presents greater dewatering problems.

Chemical sludge is produced by processes involving chemical coagulation or chemical precipitation. In conventional wastewater treatment, such processes are seldomly used. Nevertheless, they appear to offer the most economical means of removing phosphorus from wastewater, and for the treatment of intermittent flows such as combined sewage overflows. Such sludges comprise mainly the reaction products of the added chemicals and the impurities to be removed. When chemicals are applied to raw sewage, the sludge produced is a mixture of chemical sludge and organic raw sludge. If the chemical is dosed following full secondary treatment, there will be relatively little organic contamination of the sludge.

Digested sludge is the product of either aerobic or anaerobic digestion, and is a well-stabilised material capable of being dewatered on open drying beds without severe odour problems. Well-stabilised anaerobically digested sludge has a black appearance with a tarry odour; it is non-putrescible, and is therefore no longer attractive to flies and other insects.

23.3 Characterisation of wastewater sludges

The rational selection of sludge treatment and disposal units requires the definition of both the volume and the characteristics of the sludge to be treated.

The operations involved can be categorised under the general headings of conditioning, thickening, stabilisation and dewatering with the associated processes of storage, transport and ultimate disposal.

The overall aim of municipal wastewater treatment is to concentrate solids, which comprise $\leq 0.05\%$ of the mass of the raw wastewater, in sludges having gross solids contents generally in the range of 1–10%.

The total solids residue is the variable that measures the gross solids content. It is determined by evaporating the water content of a measured amount of sludge, weighing the residue and expressing it as a percentage of the original wet sludge weight. The sludge moisture content is an alternative variable, commonly quoted as a measure of the gross sludge composition.

The volatile solids content is measured as the weight loss on ignition of the dried sludge solids obtained from the total solids (TS) residue test at a standard temperature (usually 600°C). It is a measure of the organic content of the sludge. It is therefore related to the possible reduction in the sludge mass by incineration processes, and it gives an indication of the degree of stabilisation, which could be achieved by biological treatment processes. The volatile solids content is usually quoted as a percentage of the TS residue.

The solids content remaining after ignition (ash) is termed the fixed residue, and defines the weight of inorganic matter in the sludge, and thus the minimum weight of solids, which would remain for ultimate disposal after incineration.

A more detailed chemical analyses of the solids component is required where disposal methods such as composting and land application are contemplated. Such analyses should involve the determination of nutrients such as carbon, nitrogen, phosphorus and potassium.

The most important characteristics of sludges are their organic and moisture contents. A high organic content in an unstabilised sludge indicates that the sludge will continue to degrade, and is therefore likely to present health, odour and rodent problems. Table 23.1 summarises the typical moisture and organic content ranges of several sludges.

Table 23.1 Moisture and organic content of sludges

Sludge type	Moisture content (% by weight)	Organic content (% dry weight)
Primary	93–97	48–80
Activated	98–99.5	65–75
Digested	96–99	30–60
Humus	94–99	65–75
High rate plastic media	92–97	65–75

23.4 Volume of sludge

Since sludges commonly contain 1–10% solids (by weight), their major component is therefore water. Moreover, since sludge solids (i.e. biosolids in the USA)
are of similar density to water, the water content accounts for most of the volume
of wet sludges. The sludge moisture content is therefore the single variable, which
has the greatest effect on the volume of sludge to be processed at a given plant.

Furthermore, removing water from sludge with a low solids content affords
a considerable reduction in volume. Doubling the solids content from 1 to 2%,
usually halves the volume of the wet sludge.

23.5 Tests for dewatering of sludges

Wherever sludges have to be disposed of in restricted land areas or transported over
long distances for ultimate disposal, some form of volume reduction is usually
necessary. Sludge dewatering is an effective method of volume reduction in such
cases. It is also an essential pre-treatment method, where incineration is required.

Dewatering processes in common use such as pressure filters, vacuum filters
and centrifuges require for their design some measure of the sludge dewatering
characteristics. Two 'competing' methods are used to measure the ease of dewatering: specific resistance and capillary suction time (CST). The latter can be
measured by CST measurement equipment produced by Triton Electronics Ltd.
(England).

23.6 Sludge treatment and disposal objectives and methods

The main objectives of sludge treatment and disposal are as follows:

- Stabilisation of the organic matter contained in the sludge;
- Reduction in the volume of sludge for disposal by removing some of the water;
- Destruction of pathogens;
- Collection of by-products, which may be used or sold to off-set some of the
  costs of sludge treatment; and
- Disposal of the sludge in a safe and aesthetically acceptable manner.

Unfortunately, the recycling of by-products (see above) is an 'ideal scenario'
rarely achieved in practice, except in the case of methane gas, which is produced
in anaerobic digestion. The methane is often collected and used as a fuel to provide
heat for controlling the temperature of the digesting sludge and, occasionally, for
driving dual fuel engines, which may be used to generate power for the treatment
plant. The production of a compost and the use of sludges for agricultural purposes
can be viewed as waste recycling.

Sludge treatment and disposal at any particular location may comprise any or all of the following steps:

(1) Concentration: reduction in the volume of sludge to be treated by encouraging the sludge to compact to a higher solids content.
(2) Treatment to stabilise organic matter: destruction of pathogens and/or yield of by-products.
(3) Dewatering and drying: removal of water, thus reducing the sludge volume. Sludges with ≤80% moisture content are usually spadeable.
(4) Disposal: the only places where sludges can be disposed of are into the air, onto land or into water. Whether or not the impact on the receiving environment is legally, aesthetically and ecologically acceptable, depends on both the degree of treatment provided and the method of dispersing the sludge into the environment.

In comparing the various alternative methods of sludge treatment and disposal, it is important to consider all steps necessary for the final disposal of the treated sludge. An overview of the majority of available processes is shown in Table 23.2. Not all of these methods or types of methods would be practised at any one site, and not all of the methods are necessarily compatible. For each site, the range of options requires consideration.

The most commonly used method of sludge stabilisation is anaerobic digestion. Digested sludges are suitable for direct disposal, or they can be further treated, and subsequently handled without odour or hygiene problems. The principles of anaerobic digestion are set out in earlier chapters and should be incorporated within a sludge-treatment scheme. The sludge produced is amenable to dewatering. It has a reduced and stabilised organic content, and it is hygienic. However, sludge still requires final disposal.

Digested sludges are often further treated and dewatered in drying beds or in lagoons. Mechanical dewatering further reduces sludge volumes, and is a necessity for incineration. Ultimate disposal of any residue or sludge is to land or sea, though some national and international guidelines have stopped these options.

Table 23.2 Treatment and disposal options for wastewater sludges

Thickening	Stabilisation	Dewatering	Partial disposal	Ultimate disposal
Gravity	Anaerobic digestion	Drying beds	Incineration	Sanitary landfill
Flotation	Aerobic digestion	Filter press	Pyrolysis	Crop land
Centrifuge	Lagooning	Centrifuge	Wet air oxidation	Ocean
Elutriation	Heat treatment	Vacuum filter; belt press; and lagooning	Composting	
Constructed wetland (drying bed)	Constructed wetland	Constructed wetland		

If non-digested sludges are being handled, particular attention has to be paid to odour, insects and rodent infestation, and hygiene. Subsequent sections deal with particular processes in greater detail.

23.7 Treatment processes

23.7.1 Lagoons

The most common treatment process is anaerobic digestion, and the digested sludge can be further treated in a lagoon. Aerobic digestion is employed particularly at small treatment plants.

This process includes a mixture of cold digestion, air drying and gravity thickening. Sludge lagoons are usually designed on a simple 'volume per capita' basis (0.2–0.5 m^3/cap) for an expected service between 7 and 15 a before desludging is required. A water depth between 3 and 5 m is provided, and a freeboard of ≥ 1 m is required.

Internal slopes at the sides should have a width-to-height ratio of 3:1. Moreover, the sides should be lined with impervious material above, and to about 1 m below the top water level in the lagoon. Inlet arrangements should ensure even distribution of the digested sludge flow over the whole area of the lagoon floor. An outlet weir should be provided to draw off displaced water, which should then be returned to the treatment plant inlet. It is common practice to construct two lagoons; one is filled first and can be emptied while the second is filling up. The lagoons should not pollute the groundwater. Therefore, they require lining in some locations.

The sludges removed from a lagoon vary in solids content. Depending on the influent sludge, the lower compacted layers can be of 20% solids content, while surface layers contain only a few percent of solids. The sludges removed from the lagoon require final disposal.

23.7.2 Aerobic digestion

Sludges can be stabilised by an aerobic process. They are maintained in an aerobic condition by external aeration, through which organic material is oxidised. This process is widely used for sludges from small extended aeration AS plants. An aerobic digester has modest construction requirements when compared to an anaerobic digester, because it does not require gas collection. The supernatant is well oxidised, so that any return flows do not impose heavy loads on the rest of the treatment plant. The loss of methane, if compared with anaerobic digestion, is not always significant when the capital investment and running costs for using the methane are taken into account.

23.7.3 Other treatment methods

Several other treatment methods have been proposed, or are used in restricted circumstances. These include freezing and thawing of sludge to improve its dewatering characteristics. A low-pressure coolant such as butane can be employed, but it tends to dissolve greases and oils from the sludge, reducing its useful life. The fragile nature of the frozen and thawed sludge particles, together with cost, has prevented widespread use of this technique. Where freezing and thawing can be provided naturally (e.g., storage through several cold winters), the method is very effective.

Heat treatment processes such as composting can also be considered as sludge treatment processes, but are usually considered to be partial disposal methods. Other processes such as ultrasonic treatment have only restricted use due to operational problems and relatively high costs.

23.8 Thickening and dewatering of sludges

23.8.1 Chemical conditioning

The difference between thickening and dewatering methods is arbitrary. Thickening can be considered to produce a sludge of $\leq 10\%$ solids content, and tends to precede a sludge treatment process. Dewatering produces a sludge of $>10\%$ solids content, and is used after a sludge treatment process. Similar techniques can be used for thickening and dewatering. A centrifuge operated with a high throughput produces a thickened sludge, but can also be used to produce sludges of 20% solids content. Both thickening and dewatering methods can be improved by the addition of coagulants or conditioning agents.

The addition of certain chemicals increases the rate of water loss by sludge. Chemical conditioning is sometimes used in conjunction with drying beds and gravity thickening, and is a prerequisite for other processes such as filter presses, vacuum filters and centrifuges. The common conditioners are iron slats ($FeCl_3$, $FeSO_4Cl$ and $Fe(SO_4)_3$), alum ($Al_2(SO_4)_3 \cdot 18H_2O$), lime ($CaO$ and $Ca(OH)_2$) and polyelectrolytes. All these conditioners act by binding sludge particles together and permitting water to run out. Combinations of inorganic salts and polyelectrolytes can also be effective.

Lime is used to condition primary and digested sludges prior to filter pressing or vacuum filtration; it is added as 10–25% of the dry sludge solids. Applying lime to primary sludges suppresses odour formation. Iron salts can be expensive and difficult to obtain. Alum is also used prior to vacuum and pressure filtration, and drying beds. Polyelectrolytes appear to make the sludge floc stronger, and less prone to collapse during mechanical treatment. Polyelectrolytes are widely used in association with centrifugation.

23.8.2 Air drying

Sludges can be air-dried in drained drying beds, but the constraints imposed by odour control, limit this method to digested sludges. Rectangular beds are usually constructed in brick or concrete with a suitable underdrain system overlaid with sand, clinker or both materials. The beds should have a gradient of 1 in 200 to facilitate drainage. Design should promote access by foot or mechanical equipment, and provide adequate underflow. At least four drying beds, and preferably more, are required to allow a suitable drying interval between sludge applications to each bed.

Sludge is run to a depth of approximately 300 mm. After a few days, excess liquor can be removed, and returned to the treatment plant. The sludge can be lifted at approximately 70–75% moisture content when it is spadeable, or preferably at 0–55% moisture content, when it appears to be solid.

The size of a drying bed depends upon the local climate and sludge characteristics, since water is lost by evaporation and percolation. Design figures between 0.1 and 0.5 m^2/cap provide a suitable guide.

Moreover, air drying can be combined with constructed wetland drying beds. However, this is only suitable, if evapotranspiration rates are high.

23.8.3 Gravity thickening

Waste sludge can be thickened by allowing it to resettle. The general principles of settling apply to this method, with the particles undergoing hindered settlement while the bottom sludges are compressed.

Surface loadings for this process are usually between 300 and 500 m^3/m^2/d. Primary sludges can be thickened at 150 kg/m^2/d and give thickened sludge of 10% solids content. Activated sludge loads are approximately 25 kg/m^2/d, and subsequently yield sludges with 25% solids. This process is a useful pre-treatment method prior to their mechanical methods of dewatering. The effectiveness of gravity thickening can be increased by the addition of chemical conditioners. Care has to be taken during the operation of gravity thickening to minimise odours. If an undigested sludge is held for several hours, it becomes anaerobic, and hence very offensive.

23.8.4 Other methods

Washing sludge with water or effluent is known as elutration. It removes ammonia compounds, which would otherwise interfere with coagulants as well as very fine suspended matter, which is difficult to dewater. Elutration is most effective if carried out as several washings, and it is often used for primary and digested sludges prior to chemical conditioning and filtration.

Pressure filtration is a batch process in which conditioned sludge is pumped slowly with increasing pressure between filter plates, which support cloths to retain the solids and permit the passage of liquid. The pressure applied reaches approximately 700 kPa, the resultant cake is between 20 and 40 mm thick, and 40% dry solids are achieved. The process is intermittent. Modern presses have minimised the labour requirements for desludging.

In a vacuum filtration process, a continuous filter septum is partially immersed in the conditioned sludge, and is slowly rotated through it. A vacuum of approximately 85 kPa is applied inside the drum, and so liquid (i.e. water) is drawn into the drum. A cake of dewatered sludge accumulates on the outside of the drum to a thickness of approximately 50 mm. The cake is removed by scrapers. By this method, primary and digested sludges can be concentrated to 32% solids, and activated sludges to 25% solids.

With the centrifuge method, conditioned sludge is added to a rapidly (>30 rev/s) rotating bowl. The rotation separates water and sludge. Sludge solids concentrate at the outer side of the bowl, where they are removed by a screw conveyor or scroll. The sludge solids are removed at one end of the bowl, and the liquid at the opposite end. The method is compact, but requires careful control of the process variables. Centrifuges will concentrate sludges to 20% solids, but they are more commonly employed to achieve solids concentrations of approximately 10%.

Several other methods are employed to dewater sludges:

- Air flotation, which uses fine air bubbles to carry sludge solids to the surface.
- Filter belts, which compress sludge between two endless belts.
- Heat treatment processes, in which the sludge is heated to a high temperature (180–250°C) under pressure.
- Wetland systems are used to dry sludge via high evapotranspiration rates (efficient in warm climate only).
- Several hybrid methods are available; they include various combinations of methods covered above.

23.9 Partial disposal

23.9.1 Incineration

Sludges can be burnt to produce ash, which contains very little water and very little organic matter. The sludge is therefore reduced to the non-volatile fraction. A sludge containing 30% solids, of which 50% are volatile, would reduce to approximately 15% of the original wet sludge volume. Stringent controls to prevent secondary pollution are required on any gaseous or liquid discharge from incinerators.

Sludge solids have calorific values similar to conventional fossil fuels (e.g., coal and oil: 20 000–50 000 kJ/kg). Thus, dry sludges can be burnt with no additional fuel consumption. In practice, the fuel value is reduced considerably by the moisture in the sludge, so that effective dewatering is necessary prior to incineration. Designs are usually based on the production of sufficient heat to evaporate the associated water from the sludge. Sludges can be incinerated with municipal refuse.

The two types of furnace employed are the multiple hearth and the fluidised bed. Multiple hearth furnaces consist of a series of floors in a cylindrical tower; the cake is introduced at the top, and it gradually falls to the lower floors. The material is moved over floors by rabble arms. The major combustion occurs at lower levels, and heat from these levels dries out the sludge at the upper levels. In comparison, fluidised bed furnaces have a cylindrical chamber, which contains approximately 1.0 m of sand on a heat-resistant steel grid. The bed is fluidised by the injection of compressed air, and as sludge is injected into the sand under pressure through an air-cooled lance, water evaporates and the organic material burns.

23.9.2 Pyrolysis

Pyrolysis is combustion under conditions of reduced oxygen; complete oxidation is not possible. The partially combusted gaseous products from the low temperature-regulated (600°C) pyrolysis unit is passed to the high temperature-regulated (2000°C) furnace, where complete oxidation is carried out. This sequence reduces the formation of large solids (clinker) during combustion, and ensures that the gases released to the atmosphere are fully oxidised. As with incineration, the low energy value of wet sludges makes it necessary to dewater the sludges before pyrolysis, and, usually, to provide additional fuel.

An economic source of fuel is provided by municipal refuse, which can be classified to separate non-combustible materials (such as metals and glass) from the combustible fraction (such as paper and plastics). Shredded municipal refuse gives a high energy material, often referred to as refuse derived fuel. Pyrolysis of municipal refuse derived fuel and dewatered municipal wastewater sludge can therefore provide a system, which uses the properties of these two waste products, and facilitates their disposal.

23.9.3 Composting

Dewatered sludge can be composted with household refuse to produce an agricultural or horticultural compost. Non-degradable materials such as rags, metals and glass are removed from the refuse before composting. This method produces a commercial product, which has been well accepted in many communities. The process appears to be most suitable for communities in which the industrial input into garbage and wastewater is not significant.

In composting, approximately between 20 and 30% of the volatile solids are converted to carbon dioxide and water. As the organic matter in the sludge decomposes, the compost heats to temperatures in the pasteurisation range between 50 and 70°C, and enteric pathogenic organisms are destroyed.

The two most commonly used composting methods are the aerated static pile and the windrow process. The aerated static pile system consists of a grid of aeration or exhaust pipes, over which a mixture of dewatered sludge and bulking agent is placed.

In a typical static pile, the bulking agent consists of wood chips, which are mixed with the dewatered sludge. Material is composted between 21 and 28 d, and is typically cured for >30 d. Pile heights are between 2 and 2.5 m. A layer of screened compost is often placed on the top of the pile for insulation. Disposable corrugated plastic drainage pipe is commonly used for air supply. Each pile is recommended to have an individual blower for more effective aeration control. Screening of the cured compost is usually done to reduce the quantity of the end product requiring ultimate disposal, and to recover the bulking agent. For improved odour and process control, many new facilities cover or enclose all or significant portions of the system.

In the windrow process, the mixing and screening operations are similar to those for the aerated static pile system. Windrows are constructed with a height between 1 and 2 m, and corresponding measurements at the base between 2 and 4.3 m. The rows are turned and mixed periodically during the composting period. Under typical operational conditions, the windrows are turned a minimum of five times while the temperature is maintained at or above 55°C. Turning of the windrows is often accompanied by the release of offensive odours. The composting period is between 21 and 28 d. In recent years, specialised equipment has been developed to mix the sludge with bulking agents, and to turn the composting windrows. Some windrow operations are covered or enclosed (similar to aerated static piles).

23.10 Land dumping and passive treatment

Sludges can be disposed of at a sanitary landfill site, provided that this does not cause contamination of surface waters or groundwaters. The high water content of sludges usually makes it uneconomic to transport wet sludge to a landfill site, so sludges are dewatered before disposal. At small works, the sludge can be buried at or near the works. The disposal practice is labour intensive in that a trench is dug, sludge added and the area is covered again with top soil.

the sediments. Roots propagate through the soil pushing into pore space and moving soil grains. This is possible in fine-grained media — either soils or fine gravels. Roots typically die with a turnover of a few years leaving a fibrous mat of root litter.

In some cases, the decaying rhizome offers new pore space for water and gas movement. The root mat (live and dead) provides strength and stability to organic soils. *Phragmites, Typha, Juncus* and other swamp plants are widely used within treatment wetlands in Europe and Northern America (Kadlec and Knight, 1996; Mungur et al., 1997).

Metals occur in soluble or particulate forms. Heavy metals such as lead and copper are most bio-available when they are soluble; either as ionic or weakly complex forms (Cooper et al., 1996). Many trace metals are commonly found in sulphide ores; e.g., galena (PbS) and chalcopyrite ($CuFeS_2$). Metal concentrations are often high in mining wastes with associated phyto-toxic effects. Lead concentrations above 20 000 mg/kg are not uncommon (Ross, 1994).

Plant root absorption of heavy metals can be of either passive (non-metabolic) or active (metabolic) uptake in nature. Non-metabolic uptake diffuses ions in the soil solution into the outer root structures (endo-dermis). In comparison, active uptake takes place against an ionic concentration gradient, but requires metabolic energy to proceed (Alloway, 1995). For example, the uptake of lead is commonly considered to be passive, while copper is thought to be taken up either by active or by a combination of both active and passive processes (Alloway, 1995).

A diverse microbial community including a variety of bacteria, fungi, algae and protozoa is present in both the aerobic and anaerobic zones of wetlands (Kadlec and Knight, 1996). Moreover, most wetland species have a symbiotic relationship of their roots with specialized fungi, known as mycorrihzas ('fungus roots'). In the presence of mycorrhizae, an increased uptake of copper has been observed in soils low in nutrients (Wild, 1993).

The aim of this experimental study is to characterize the function of biomass within mature experimental wetland filters with different design characteristics including the selection of appropriate macrophytes and aggregates. The removal efficiencies of mature experimental constructed wetlands treating urban water receiving high loads of lead and copper will be assessed. An overall water quality assessment focusing on physical and chemical variables as well as bio-indicators will be undertaken. Moreover, statistical relationships between water quality variables will be identified to reduce sample costs and effort.

24.3 Materials and methods

24.3.1 Experimental plan and limitations

The experimental plan focuses on the water quality investigation of six different experimental wetland filters receiving urban water contaminated with lead and copper. The metal concentrations were increased after one year of operation in order to assess the associated impact on the outflow water quality and biomass within the filters.

This was based on applied engineering science principles; therefore, no replicate filters were operated. Moreover, the operation of two or three replicates for each filter (up to 24 filters in total) would have been impractical and unrealistic on a small pilot plant scale as the one presented in this chapter.

Results are not represented in terms of mass balance equations because most experimental data vary greatly with season. Moreover, the focus of this study is on the water quality that is traditionally not expressed in terms of mass balance but concentration units.

24.3.2 Filter media composition

Wetland habitats were simulated on a laboratory scale with experimental filters of the following dimensions: height (58 cm), upper diameter (40 cm) and lower diameter (32 cm). The empty filter volume was 59 l. Six packing arrangements of filter media and plant roots were used in six vertical-flow wetland filters (Table 24.1).

Filter 1 is similar to a slow sand filter. In comparison, Filters 2 and 3 are typical reed bed filters. However, Filters 4 to 6 are predominantly of academic interest, because they contain different absorption media.

Additional adsorption media were used, because they are associated with high heavy metal removal potentials (Scholz and Martin, 1998a,b). The objective was to study the influence of different compositions of aggregates on the overall treatment efficiency.

Table 24.1 Packing order arrangements of vertical-flow filters simulating wetlands

Height (cm)	Filter 1	Filter 2	Filter 3	Filter 4	Filter 5	Filter 6
49–58 (top)	Water	Water	Water	Water	Water	Water
47–48 (top)	Sand	Sand + P	Sand + P + T	GAC + P + T	GAC + P + T	Charcoal + P + T + Fertilizer
41–46 (top)	Sand	Sand + P	Sand + P + T	GAC + P + T	GAC + P + T	GAC + P + T
37–40 (top)	Sand	Sand + P	Sand + P + T	Sand + P + T	Filtralite + P + T	Filtralite + P + T
35–36 (top)	Sand	Sand	Sand + T	Sand + T	Filtralite + T	Filtralite + T
33–34 (top)	Gravel	Gravel	Gravel + T	Sand + T	Sand + T	Sand + T
25–32 (top)	Gravel	Gravel	Gravel	Sand	Sand	Sand
21–24 (top)	Gravel	Gravel	Gravel	Gravel	Sand	Sand
0–20 (bottom)	Gravel	Gravel	Gravel	Gravel	Gravel	Gravel

GAC = granular activated carbon; P = Common reed or *Phragmites australis* (Cav.) Trin. Ex Steud; T = Cattail or *Typha latifolia* L.

24.3.3 Environmental conditions and operation

The experiment was performed in the center of a large unheated laboratory hall (windows at two sides and at the top) that was subject to direct and indirect variable natural (day-length) and artificial light. The system was operated in batch flow mode to save pumping costs. The inflow water was taken from the Westbrook (a local brick-lined urban stream), which is polluted by urban runoff and sewage from combined sewer overflows during and after storms. The pH of the raw stream water was slightly acidic. The experiment was split into two operational stages; both stages lasted one year each in order to assess the system performance for all seasons; 21/01/00 to 20/01/01, and 21/01/01 to 20/01/02.

Lead and copper were selected as pollutants, because they are both not easily bio-available. Lead and copper sulphate were added to the inflow water to give total concentrations of dissolved lead and copper of approximately 1.30 and 0.98 mg/l, respectively, for the first, and 2.98 and 1.93 mg/l, respectively, for the second year. The magnitude of lead and copper values is comparable to figures reported for (pH adjusted) mine wastewater and urban water heavily contaminated with heavy metals (Cooper et al., 1996; Hammer, 1989; Kadlec and Knight, 1996).

Natural background concentrations of the Westbrook water (before contamination with heavy metals in the laboratory) were approximately 0.02 mg/l for lead and 0.06 mg/l for copper. These background concentrations are very low in comparison to the values measured after artificial contamination.

The calculated theoretical hydraulic load per filter was approximately 1.8 cm/d for the first and 4.2 cm/d for the second stage. The hydraulic load is based on the mean inflow volume (m^3) of water that flows through the cross section (m^2) of a filter per day (d). Approximately 40% of the filter water was drained and replaced with 6 l on Mondays, Wednesdays and Fridays for the first stage and 7 l contaminated inflow water on each working day for the second stage of the experiment.

Calculated mass load rates for lead and copper were 24.3 and 18.0 mg/m^2/d, respectively, for the first, and 118.8 and 77.0 mg/m^2/d, respectively, for the second stage of the experiment. The mass load rate is based on the metal inflow concentrations (mg/m^3) associated with the mean daily inflow volume (m^3) distributed over the cross section (m^2) of each wetland filter. After one year of operation, high metal loads were applied to stress the biomass even further and to explore the limits of adequate system performance.

The water level was lowered carefully within two minutes until only 0.5 cm of water remained on top of the submerged litter. The purpose of this procedure was not to destroy the anoxic biofilm structure. The inflow was distributed slowly and evenly into the schmutzdecke (i.e. layer of dirt containing suspended solids and biomass) and litter zone of each filter. The effluent point was located at the center of the bottom of the filter.

The balanced controlled release N–P–K fertilizer Miracle-Gro (formerly Osmocote, produced by Scot Europe B. V., The Netherlands) was added to the top of Filter 6 to stimulate the growth of macrophytes. Approximately 50 g of the controlled release Miracle-Gro pellets was added in January, April, July and October. The applied fertilizer load was low according to manufacturer's guidelines. The initial load was therefore doubled after one year of operation to enhance further macrophyte and biomass growth in Filter 6 only.

24.3.4 Analytical procedures including metal determination

For heavy metal determination, sample water (5 cm^3) from the inflow (before adding metals), outflow and from the water layer on top of the litter zone (sample point selected at random) was directly analyzed with a Palintest Scanning Analyzer SA-1000 approximately twice a month. The performance range of this system was between 0.002 and 0.100 mg/l ($\pm5\%^{CV}$ at 0.015 mg/l) for lead and between 0.070 and 2.000 mg/l ($\pm10\%^{CV}$ at 0.300 mg/l) for copper. Pre-calibrated electrodes (SE-1), prepared by the manufacturer, were used. Mean values based on at least two readings per sample were calculated. Sample water dilutions were made, if appropriate.

In October 2001, metal concentrations within the schmutzdecke, litter, macrophytes and liquid samples were analyzed by Atomic Absorption Spectroscopy in order to increase sample accuracy. Samples were taken randomly, and two or three sample replicates were considered. A Perkin Elmer 1100 series Atomic Absorption Spectrophotometer was used for metal determination with an accuracy of 0.19 mg/l at 283.3 nm for lead and 0.032 mg/l at 324.9 nm for copper (Perkin Elmer, 1982).

Macrophyte and aggregate (including microbial biomass and litter) samples were taken randomly. Different sample types for macrophytes (leaf, stems and rhizome) and depth profiles for aggregates were analyzed for lead and copper concentrations to determine the effective heavy metal load removal capacity from the contaminated waters by the separate components of the constructed wetland filters.

Water samples were analyzed directly, but solid materials such as rhizomes were dissolved before the corresponding solution was analyzed. Nitric acid was considered to be suitable for flame and non-flame atomization. In comparison, sulphuric acid can interfere with flame atomization. Perchloric acid and hydrochloric acid were not used for samples containing lead because volatile chlorides could be formed. These chlorides can be lost before flame atomization and subsequent analysis (Tan, 1996).

It was necessary to use an oxidizing acid such as nitric acid (alternatively, perchloric acid) in combination with hydrofluoric acid, because decomposed organic material (e.g., aquatic plant litter) may contain sulphides (Alloway, 1995). Due to health and safety considerations regarding the use of hydrofluoric acid and the requirement for lead determination, a mixed acid digestion of concentrated nitric acid (10 ml) and sulphuric acid (1 ml) was used to maximize biomass dissolution with minimal risk to health and safety, and experimental accuracy.

During the atomic absorption analysis, the spectrophotometer was calibrated using 0.0, 0.5, 1.0, 1.5, 2.0, 5.0, 7.5, 10.0, 15.0 and 20.0 ppm (i.e. mg/l for liquid samples) standards for lead analysis and, 0.00, 0.25, 0.50, 0.75, 1.00, 1.50, 2.00, 4.00, 5.00 and 8.00 ppm standards for copper analysis. The calibration curves were calculated by least squares linear regression analysis (coefficient of determination $r^2 > 0.97$). The metal concentrations for the samples were then calculated with the help of the corresponding calibration curves.

A Hanna HI 8519N micro-processor bench-top pH meter (Orion Model 420 used as a replacement after 14/09/2001), a Hanna HI 9142 portable waterproof dissolved oxygen (DO) meter, a Hanna C 102 turbidity meter and a Hanna HI 9033 conductivity meter were used to determine pH, DO, turbidity and conductivity, respectively. An ORP HI 98201 redox meter with a platinum tip electrode HI 73201 was used.

Measurement units used are listed in Table 24.2. Samples were analyzed on Mondays, Wednesdays and Fridays of most working weeks. No replicates were taken. All analytical procedures were performed according to the standard methods (Clesceri et al., 1998) except counting of microorganisms (explained below).

24.3.5 Micro-biological examinations

Samples were taken twice a week for analysis of micro-organisms. Five water samples (5 ml each) were taken randomly from the inflow water tank, outflow water tank and the top layer of filter media (mixed with plant litter). The samples from each sample set were mixed with each other to reduce sample bias, and to increase sample accuracy and validity of the subsequent analysis of sub-samples. The sample regime is based on the assumption that all water sources have a homogenous water quality at the time of sampling. Immediately after mixing, microorganisms including protozoa and algae groups were counted in a first sub-sample (two replicates) of each filter by using a Fuchs-Rosenthal Haemacytometer (counting chamber: $0.20 \times 0.25 \times 0.25$ mm) and a Wang Biomedical Research Microscope 6000 (bright field and phase-contrast).

During counting, one micro-organism counted represented 313 organisms per ml. Protozoa were identified based on their mode of movement (i.e. life action; e.g., movement with flagella or cilia) and morphology according to the methodology applied for the general subdivision of protozoa in British sewage treatment processes (Scholz and Martin, 1998a,b). The research was focused on microbiological parameter that can be interpreted as bio-indicators for the water quality (e.g., ciliated protozoa) or essential to assess filter clogging (e.g., diatoms).

Table 24.2 First (21/01/00–20/01/01) and second (21/01/01–20/01/02) year of the experiment: inflow Westbrook water quality parameter after contamination with lead and copper sulphate

	Variable	Unit	Sample number	Mean	Median	Minimum	Maximum	SD
First year	Lead	mg/l	23	1.30	1.28	≈1.0	≈1.5	n/a
	Copper	mg/l	23	0.98	0.99	≈0.5	≈1.5	n/a
	BOD	mg/l	127	2.7	1.9	0.1	11.3	2.4
	SS	mg/l	77	15.8	12.0	<0.1	609.0	15.4
	Total solids	mg/l	96	470	472	114	1048	147
	Turbidity	NTU	91	8.9	6.8	0.6	49.5	8.0
	DO	mg/l	139	8.2	8.1	5.8	10.4	0.6
	pH	–	135	7.69	7.71	6.96	8.07	0.20
	Redox potential	mV	51	284	302	76	423	70
	Conductivity	S	74	674	631	271	997	186
	T (at source)	°C	114	11	12	7	26	6
	T (laboratory)	°C	115	20	20	10	28	3
Second year	Lead	mg/l	18	2.98	2.98	2.5	3.5	n/a
	Copper	mg/l	18	1.93	1.93	1.5	2.5	n/a
	BOD	mg/l	139	2.8	2.1	<0.1	10.0	2.45
	SS	mg/l	77	19.4	18.0	<0.1	242.0	12.1
	Total solids	mg/l	56	404	430	20	846	170
	Turbidity	NTU	104	11.6	8.2	<0.01	80.0	12.5
	DO	mg/l	140	8.2	8.3	3.9	10.2	0.9
	pH	–	123	7.47	7.48	6.50	8.06	0.26
	Redox potential	mV	105	195	195	98	310	42
	Conductivity	S	90	657	675	143	836	118
	T (at source)	°C	163	9	10	2	24	6
	T (laboratory)	°C	147	18	18	11	26	3

SD = standard deviation; BOD = five-day biochemical oxygen demand; SS = suspended solids; DO = dissolved oxygen; n/a = not available (The SD could not be estimated accurately, because most estimated sample values were below or above the detection limit of the mostly diluted sample; see also the section on analytical procedures including metal detection.); T = temperature.

24.3.6 Statistics

The Microsoft Excel correlation analysis, coefficient of determination and trend-line fitting tools were used for practical statistical data analysis. Trendlines to microbiological data (see below) were not fitted to show the variance associated with biological data. Furthermore, all second-order polynomials were fitted by calculating the least squares fits through points representing corresponding pairs.

24.4 Results and Discussion

24.4.1 Comparison of treatment efficiency

Table 24.2 summarises the inflow water quality after the addition of dissolved lead and copper sulphate to natural Westbrook water. The standard deviations for all inflow variables (except for DO; the stream is fast flowing) were elevated (Table 24.2) due to natural and artificial water quality variations of the Westbrook. The water quality of the stream in terms of BOD and SS was sometimes low due to irregular sewage input via a combined sewer overflow structure located upstream of the abstraction site. The correlation between SS and turbidity was low because of the presence of large inorganic solids (mostly fine sand).

After two years of operation, the overall performance in terms of copper, lead, BOD and turbidity removal was high for most filters (Table 24.3), if compared to published data for similar reed beds (Hammer, 1989; Kadlec and Knight, 1996). However, the reduction in SS was negligible. An analysis of variance (Fowler and Cohen, 1998) indicated that reductions of lead and copper were similar ($P < 0.05$).

It has been suggested that mature biomass, in contrast to aggregates with high adsorption capacities, is responsible for the high overall filtration performances (Cooper et al., 1996; Scholz and Martin, 1998a,b). However, it is difficult to classify objectively a biological system as mature. Nevertheless, there was no breakthrough of metals after approximately nine weeks. Furthermore, a stable

Table 24.3 First (21/01/00–20/01/01) and second (21/01/01–20/01/02) year of the experiment: relative reduction (%) of outflow variables

Variables	Change (%) per wetland filter[a]					
	1	2	3	4	5	6
First year						
Lead[b]	>98	>98	>98	>99	>99	>99
Copper[b]	>83	>89	>90	>93	>92	>93
Five-day biochemical oxygen demand	77	79	70	75	80	72
Suspended solids	50	37	40	46	48	25
Total solids	4	−7	−7	−7	−4	−24
Turbidity	94	93	88	87	96	94
Dissolved oxygen	54	50	69	68	61	68
pH	1	7	8	7	6	8
Redox potential	−2	−7	14	10	6	3
Conductivity	−2	−15	−18	−20	−15	−36
Second year						
Lead[b]	>100	>99	>99	>99	>100	>100
Copper[b]	>96	>93	>96	>97	>93	>97
Five-day biochemical oxygen demand	74	77	70	72	76	62
Suspended solids	53	44	49	39	36	24
Total solids	5	−3	2	−1	−4	−27
Turbidity	87	86	86	87	78	86
Dissolved oxygen	53	49	72	74	65	75
pH	0	7	7	7	7	11
Redox potential	−4	−11	−3	24	15	4
Conductivity	1	−5	−2	−6	−5	−25

[a]Change (%) = (in-out)×100(%)/in, where in = inflow value and out = outflow value.
[b]Some outflow concentrations were below the detection limit.

biological ecosystem in terms of the abundances of protozoa and crustaceans has been noted.

24.4.2 Water quality and macrophytes

The filters containing macrophytes (in particular, *Typha latifolia*; Filters 3–6) contributed artificially to the inflow BOD. Approximately 75% of *Typha latifolia* (infested by aphids) and 30% of *Phragmites australis* died during late autumn. Dead plant material made a fair contribution to the top BOD (measured at the top of the filter) of Filters 3–6 (data not shown). The 5-day BOD resulting from plant decay was greatest for filters containing *Typha latifolia*. The addition of fertilizer (Filter 6 only) increased the degradation rate of dead plant matter and the breakthrough of TS. Furthermore, DO was reduced.

A positive correlation was expected between decaying macrophytes (expressed as an increase in BOD) and conductivity. However, the presence of macrophytes did not increase conductivity (Fig. 24.1); e.g., Filters 1 and 2 had similar conductivity values, and temporal distributions.

Due to photosynthesis of green algae, Filter 1 had the highest daily observed DO concentrations (at the top of the filter), which were even higher than the ones for the inflow water for most days in spring and summer. Concerning Filters 3–6, the top and outflow DO concentrations were almost always <4.0 mg/l. However, the outflow DO values were slightly higher than the top concentrations (data not shown).

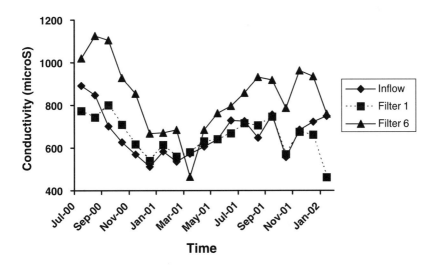

Fig. 24.1 Monthly average conductivity values (01/07/00–20/01/02) for the inflow, Filter 1 (outflow) and Filter 6 (outflow).

The differences in top DO can be explained as follows: decaying plants in the litter zone indirectly supplied ample dissolved organic matter for the active biomass. It follows that microbial respiration reduced the DO concentrations within the litter zones of Filters 2–6. Another reason for the differences in DO is the presence of green algae, which were thriving in Filter 1 during summer.

In contrast, green algae decreased in Filters 2–6 due to the shading effect from the emergent plants. As a result of the photosynthetic oxygen production by green algae, the DO and pH concentrations for Filter 1 were higher than the concentrations in Filters 2–6. Therefore, algal growth did not compensate for the oxygen utilised by microbial respiration. Furthermore, dead plants floating on the water surface of Filters 2–6 impeded oxygen dissolution in the open water zones by reducing the areas of water–oxygen interface. Moreover, the decay of macrophytes reduced the pH as indicated in Fig. 24.2.

In comparison to the inflow, concentrations of lead and copper were low within macrophytes and aggregates (Table 24.4). Because macrophytes contain low concentrations of metals at the end of the growing season, harvesting of macrophytes (i.e. mature leaves and stems only) leads only to a small reduction of the overall metal concentration within the filters. In contrast, harvesting macrophytes during the growing season, where metals are accumulated in the new stems and fresh leaves, might be more promising, but will damage the ecosystem considerably.

Relatively high concentrations of copper were present in the top aggregate layer and the litter zone (44–48 cm) of the reed bed filters. In contrast, relatively high lead concentrations occurred throughout the top part (32–48 cm) of the filter beds

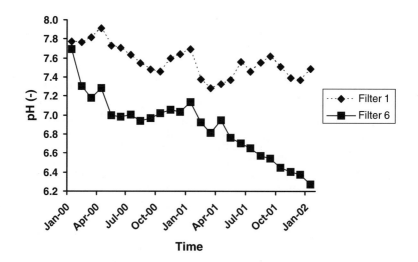

Fig. 24.2 Monthly average pH values (01/07/00–20/01/02) for Filter 1 (outflow) and Filter 6 (outflow).

(Table 24.4). Much higher metal concentrations were expected for the top of the filters compared to the inflow concentrations. A realistic assumption is that all filters have a high remaining filtration capacity and that particulate and colloidal metals integrated into the biomass and absorbed by aggregates are difficult to detect fully with the atomic absorption methodology presented above. Lead (to a lesser extent, copper) forms also inert particulate matter that breaks slowly through the filters. Different aggregate layers help to slow down this process not only by physical straining but also by complex filtration processes.

Biological processes within the process are considered to be important, but their investigation was beyond the scope of this experiment. For example, precipitation of metal oxides and sulphates may be caused by the presence of metal-oxidizing bacteria in the aerobic zone and sulphate-reducing bacteria in the anaerobic zone, respectively (IWA, 2000). However, lead and copper are not easily bio-available to aquatic plants and microbes. Furthermore, a relatively high pH value reduced the risk of metals going back into solution (Table 24.2).

Table 24.4 Atomic absorption spectroscopy results: mean concentrations (mg/kg) of lead and copper for decomposed macrophyte material and biomass located within the aggregates of different filter depths (December, 2001). Values are based on the mean of either two or three replicates. Metal concentrations exceeding the inflow concentrations are highlighted (i.e. numbers in italic)

Variable	Metal concentrations (mg/kg) within wetland Filters 1–6											
	Filter 1		Filter 2		Filter 3		Filter 4		Filter 5		Filter 6	
	Pb	Cu	Pb	Cu	Pb	Cu	Pb	Cu	Pb	Cu	Pb	Cu
Inflow water	2.98	1.93	2.98	1.93	2.98	1.93	2.98	1.93	2.98	1.93	2.98	1.93
Phragmites (leaf)	n/a	n/a	0.37	0.47	0.28	0.38	0.22	0.39	0.38	0.61	0.55	0.50
Phragmites (mature stem)	n/a	n/a	0.28	0.47	0.24	0.48	0.40	0.47	0.33	0.46	0.38	0.48
Phragmites (new stem)	n/a	n/a	0.36	0.91	0.20	0.51	0.12	0.17	0.29	1.31	*2.91*	*3.06*
Phragmites (rhizome)	n/a	n/a	0.63	0.74	*5.79*	1.03	0.12	0.19	0.59	0.72	0.29	0.38
Typha (leaf)	n/a	n/a	n/a	n/a	*6.17*	1.07	0.18	0.42	0.31	0.50	0.12	0.72
Typha (stem)	n/a	n/a	n/a	n/a	0.36	0.39	0.31	0.63	0.14	0.62	0.20	0.60
Layer height: 44–48 cm	*4.16*	*2.83*	0.28	*3.76*	0.32	*4.38*	0.71	*12.28*	1.91	*8.22*	*6.35*	*9.67*
Layer height: 40–44 cm	2.17	*1.98*	0.55	0.57	0.32	0.59	0.53	0.80	0.38	1.00	0.44	0.63
Layer height: 36–40 cm	0.67	0.78	*4.22*	0.80	*6.74*	0.12	*4.97*	0.18	*12.23*	0.16	0.33	0.32
Layer height: 32–36 cm	2.02	*2.36*	1.01	1.82	1.32	*1.99*	2.78	2.76	1.10	1.87	1.26	*1.96*
Layer height: 28–32 cm	1.00	1.82	0.90	1.76	1.04	1.84	0.95	1.79	0.92	1.78	0.88	1.76
Layer height: 24–28 cm	0.85	1.74	0.89	1.76	1.09	1.87	1.08	1.86	0.90	1.77	0.88	1.76
Fertilizer	n/a	n/a	n/a	n/a	n/a	n/a	n/a	n/a	n/a	n/a	0.33	0.12
Charcoal	n/a	n/a	n/a	n/a	n/a	n/a	n/a	n/a	n/a	n/a	0.43	0.08
GAC	n/a	n/a	n/a	n/a	n/a	n/a	0.16	0.06	0.16	0.06	0.16	0.06
Filtralite	n/a	n/a	n/a	n/a	n/a	n/a	n/a	n/a	0.16	0.21	0.16	0.22
Sand	0.41	0.10	0.41	0.10	0.41	0.10	0.41	0.10	0.41	0.10	0.41	0.10

Pb = lead; Cu = copper; n/a = not applicable; GAC = granular activated carbon.

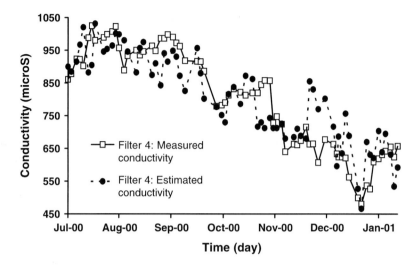

Fig. 24.6 Measured and predicted outflow conductivity for Filter 4.

24.5 Conclusions

Adsorption filter media (as replacements for aggregates such as sand and gravel) did not significantly reduce lead and copper concentrations after the establishment of an active biomass (two years old). In this study, the overall filtration performance was acceptable and similar for all experimental filters after biomass maturation. It follows that the biomass (not the aggregates and macrophytes as expected) within each filter is predominantly responsible for the removal of lead and copper. This will have implications on the future design and operation of constructed wetlands. Costs can be saved on filter material and aquatic plants.

Lead and copper accumulated predominantly within the litter zone and the rhizomes of the macrophytes (practically not accessible). It follows that costs could be saved by constructing shallow filter beds. Furthermore, only a limited load of heavy metals could be removed from the constructed reed beds by harvesting macrophyte stems and leaves. The costs associated with harvesting do not currently justify the benefit of metal removal.

The dissolved fraction of metals within the litter zone leads to a reduction in bioactivity (e.g., ciliated protozoa and zooplankton) during the second spring and summer of the experiment. The reduction of these bioindicators reflected the accumulation of toxic metals within sediments and the food chain. This has a negative environmental impact on the wildlife associated with wetlands. Furthermore, diatom numbers increased during the second stage resulting in potential filter clogging problems.

Fertiliser use in one of the filters leads to a deterioration of the water quality, particularly with respect to SS and pH. Additional nutrients lead to an artificial increase of the biodegradation rate, which is not desirable in case of treatment wetlands for metal removal. Thus, costs can be saved by not applying fertilisers.

Strong positive correlations between conductivity and other variables including temperature were identified. Some expensive and time-consuming (i.e. high sampling effort) variables such as BOD, DO and conductivity can be predicted with less expensive ones such as temperature that are easy to measure. This would result in the saving of costs and sampling effort for future constructed wetland water quality monitoring schemes.

Chapter 25

Wetland systems to control roof runoff

25.1 Summary

The purpose of this case study was to optimise design, operation and maintenance guidelines, and to assess the water treatment potential of a wetland system following 15 months of operation. The system was based on a combined silt trap, attenuation pond and vegetated infiltration basin. This combination was used as the basis for construction of a roof runoff system from a single domestic property.

Design guidelines from the UK Building Research Establishment, Construction Industry Research and Information Association, and the German Association for Water, Wastewater and Waste were tested. These design guidelines failed, because they did not consider local conditions. The infiltration function for the infiltration basin was logarithmic.

Algal control techniques were successfully applied, and treatment of roof runoff was found to be largely unnecessary for recycling purposes (e.g., watering plants). However, seasonal and diurnal variations of BOD, DO and pH were recorded.

25.2 Introduction

25.2.1 Sustainable roof runoff drainage

Considering that sustainable urban drainage system (SUDS) technology is a novel research area, definitions for the terminology used in this chapter are required. Reduction of the rate of flow through a system, which has the effect of reducing the peak flow and increasing the duration of a flow event, is defined as 'attenuation'. A 'pond' is specifically referred to as a 'wet pond', if it is a permanently wet depression designed to retain stormwater for several days, and to permit settling of SS. It follows that an 'attenuation pond' combines both the meaning of 'attenuation' and 'pond'. In comparison, an 'infiltration basin' is a dry basin designed to promote infiltration of surface water into the ground. If vegetated, a small 'infiltration basin' is often referred to as a 'dry pond'.

Conventional stormwater and urban drainage systems are designed to dispose off rainfall runoff as quickly as possible. This results in 'end of pipe' solutions that often involve the provision of large interceptor and relief sewers, huge storage tanks in downstream locations and centralised wastewater treatment facilities (Abbott and Comino-Mateos, 2001; Butler and Davies, 2000; Ellis et al., 2002).

In contrast, combined attenuation pond and infiltration basin systems can be applied as cost-effective 'end of pipe' drainage solutions for local source control; e.g., collection of roof drainage. It is often possible to divert all roof drainage for infiltration or storage, and subsequent recycling. As runoff from roofs is a major contributor to the quantity of surface water requiring disposal, this is a particularly beneficial approach, where suitable ground conditions prevail (Butler and Davies, 2000).

Furthermore, roof runoff is usually considered to be cleaner than road runoff. However, diffuse pollution can have a significant impact on the water quality of any surface runoff (Ellis et al., 2002).

25.2.2 Case study: site description

A domestic property in Sandy Lane (Bradford, West Yorkshire, England) was selected for this pilot plant case study. The study area is located approximately 1.8° west of Greenwich and 53.8° north of the Equator. The surface water (subject to disposal) came from the roofs of the house and a tandem (double) garage.

Concrete roof tiles were used for the house, and the garage roof was covered with a layer of gravel. The roof materials had not been cleaned for >5 a prior to the study (current and previous house owners, personal communications). One can assume that dirty roofs could be a source of inorganic and organic pollution, and it is likely that the roof material supports micro-habitats that include algae. However, a detailed discussion of these problems is beyond the scope of this chapter.

25.2.3 Purpose

The purpose of this chapter is not only to research a case study of a SUDS designed according to best management practice (Martin et al., 2000) but also to address the following key objectives to assess the potential for scaled-up systems:

- Identification of technical constraints associated with the design, operation and maintenance;
- Identification of the infiltration characteristics of stormwater into the vegetated infiltration basin;

- Suggestion of water quality monitoring and management strategies including algal control techniques; and
- Assessment of the water treatment potential of stormwater ponds for roof runoff.

25.3 Methods

25.3.1 Design of the study site

The pilot plant was designed considering SUDS guidelines of the BRE (1991), CIRIA (Bettes, 1996; Martin et al., 2000) and ATV-DVWK (2002). The system design allows flooding to occur only once within 10 a, which is defined as the return period. However, the German guideline recommends a design allowing for a five-year storm only (ATV-DVWK, 2002).

The system was based on a combined attenuation pond and infiltration basin design (Fig. 25.1). Runoff from the roofs of one house (tandem garage attached) was drained directly into a silt trap, which fulfilled the purpose of a small sedimentation tank.

The surface areas of both house roofs measured 29 m^2 each. The angle between each roof and the ceiling of the house was 23°. Therefore, the total theoretical horizontal area of the house roofs was 53 m^2. The roof area of the tandem garage measured 33 m^2. It follows that the total theoretical horizontal area to be drained was 86 m^2.

The distances between the nearest building (garage) and the attenuation pond and infiltration basin were 1.5 and 5.0 m, respectively (Fig. 25.1). The total length of the horizontal plastic pipe work close to the ground (mean angle of 2°) was 19.6 m. Guttering and down pipes were not included in this sum. The inside diameter of all pipes was 6.5 cm. The pipeline layout could have been optimised, but the SUDS was retrofitted in this case study, and it was therefore the aim to recycle as much of the old pipework as possible.

In the original pipeline layout dated 1972, rainwater drained into the public sewer. However, in April 2001, this layout was changed to feed a semi-natural attenuation pond structure (Fig. 25.1) with rainwater. The storage water was predominantly used for watering garden plants in summer, but there is a much greater potential for other usage; such as recycling of stormwater to flush toilets (Butler and Davies, 2000). If the attenuation pond structure overflows, the water is transferred to a vegetated infiltration basin structure (Fig. 25.1).

The traditional drains were sealed off temporarily, although the house owners preferred a system that would allow rainwater diversion into the main sewer in cases of exceptional rainfall or failure of the infiltration basin. However, neither of these events occurred during the experiment.

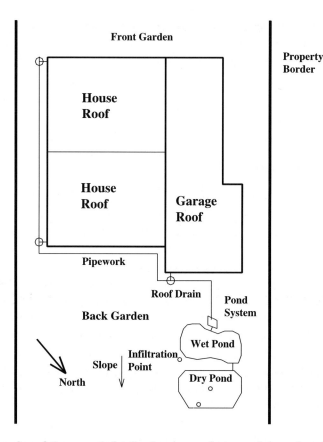

Fig. 25.1 Drawing of the case study site showing roof areas, pipe work and the combined attenuation wet and dry pond system. Runoff flows from the roof areas into the drainage pipe network, which conveys the water first into the silt trap, then into the wet pond and finally into the dry pond.

The maximum horizontal dimensions (length × width) of the silt trap, attenuation pond and infiltration basin were 0.7×0.4 m, 3.2×1.7 m and 3.7×2.5 m, respectively (Fig. 25.1). The maximum depths of the lined attenuation pond and the vegetated infiltration basin were approximately 40 cm each. All water level measurements were taken daily at a reference level point that was part of the attenuation pond outflow structure (Fig. 25.1).

The total area of the experimental attenuation pond was 5.5 m^2, when completely filled with water (Fig. 25.1). The slope ratios of the infiltration basin were 1:1.6 (towards the attenuation pond in the West), 1:1.9 (towards North and South) and 1:2.4 (towards the lower garden in the East).

The attenuation pond and infiltration basin were planted with common aquatic plants, which were collected from local natural habitats. The dominant macrophytes of the attenuation pond included Common Reed (*Phragmites australis*), Reedmace (*Typha latifolia*) and Yellow Iris (*Iris pseudochorus*). Water Starwort (*Callitriche stagnalis*) and Frogbit (*Hydrocharis morsus-ranae*) were common floating aquatic plants. Canadian Waterweed (*Elodea canadensis*) was the dominant submerged plant. Common Reed and Reedmace (both deep-rooting) were also planted in the infiltration basin to enhance infiltration properties, although this has been disputed by some findings reviewed by Brix (1997). This combination of plants was chosen, because it is typical for similar watercourses around Bradford.

Two koi fish (introduced during pond construction in April 2001), at least 25 European three-spined sticklebacks (*Gasterosteus aculeatus*; introduced in September 2001) and the common grass frog (*Rana temporaria*; approximately 20 adults and at least 2500 tadpoles during spring and summer) were at the top of the food chain in the attenuation pond.

25.3.2 Engineering methods

Two simple rain gauges were used (comprising of a measurement cylinder fed through a funnel with a diameter of 9 cm) to give a more representative estimate of rainfall. Experimental rainfall data were compared with official data (Bradford measurement station) supplied by the Meteorological Office (2002). The rain gauges were located between the attenuation pond and the infiltration basin (Fig. 25.1).

A cost-effective, non-destructive test on a small-scale was required to assess the stormwater (roof runoff) infiltration properties into the infiltration basin. This information was important for the estimation of the infiltration rates required for the design calculations (see above). Therefore, a simple infiltration device specifically designed to promote the passage of gravity-fed tap water into the natural ground was used.

Infiltration rates were determined fortnightly or monthly by measuring the actual infiltration time of 200 ml tap water through the first 4 cm of a 24-cm long drainage pipe (diameter of 6.5 cm). The pipe was buried vertically in the ground at a depth of approximately 20 cm. The infiltration test was carried out at three infiltration point stations in parallel under varying geological, biochemical and particularly meteorological conditions. The infiltration stations are indicated by small circles in Fig. 25.1.

25.3.3 Water quality analysis

Daily or weekly water quality sampling schemes were applied. Daily sampling and subsequent analysis took place at approximately 06:00 and/or 18:00 h. All analytical procedures to determine the water quality were performed according to the American standard methods (Clesceri et al., 1998) that also outline the corresponding water quality criteria. Water samples were tested for temperature (air and water), BOD, SS, TS, conductivity, turbidity, DO and pH. The 'Hanna instruments' HI 9033 conductivity meter, C 102 turbidity meter, HI 9142 DO meter and HI 8519N pH meter were used throughout the study.

Oxidised aqueous nitrogen was determined as the sum of nitrate and nitrite. Nitrate was reduced to nitrite by cadmium and determined as an azo dye at 540 nm (using a Peristorp Analytical EnviroFlow 3000 flow injection analyser) following diazotisation with sulfanilamide and subsequent coupling with N-1-naphthylethylenediamine dihydrocloride. This technique also simultaneously measured nitrite. Furthermore, aqueous ammonia reacted with hypochlorite and salicylate ions in solution in the presence of sodium nitrosopentacyanoferrate (nitroprusside), and aqueous phosphate reacted with acidic molybdate to form a phosphmolybdenum blue complex. The associated coloured complexes were measured spectrophotometrically at 655 and 882 nm, respectively, using a Bran and Luebbe Autoanalyser (Model AAIII). All analyses for nutrients were carried out in triplicate.

Vegetative, sediment and soil samples were dried at 105°C overnight in a drying oven (UM500, Memmert) prior to being ashed at 400°C for 12 h in a muffle furnace (ELF 11/14, Carbolite). Ashed samples (0.2–0.6 g) were then digested under reflux in *aqua regia* for 2 h, cooled, filtered through Whatman No. 5412 filter paper and diluted to 100 ml with de-ionised water ready for analysis.

An ICP-OES (TJA IRIS instrument manufactured by ThermoElemental, USA), at 1350 W with coolant, auxiliary and nebuliser argon gas flows of 15, 0.5 and 0.7 ml/min, respectively, and a pump flow rate of 1 ml/min, was used to screen for total concentrations of analyte elements in filtered (0.45 μm) water samples and digests.

Multi-element calibration standards in the concentration range between 0.1 and 10 mg/l were used, and the emission intensities were measured at appropriate wavelengths in nm. For all elements, analytical precision in terms of the relative standard deviation was typically between 5 and 10% for individual aliquots. Three replicates for each sample were analysed.

An ETAAS (Varian SpectrAA 400 manufactured by Varian Inc., Australia) with auto-sampler and powered by a GTA-96 graphite tube atomiser was used to analyse some of the water samples for their zinc content. A 20-μl injection volume was used for samples and standards in notched GTA partition tubes (coated). Nitrogen was applied as the carrier gas. The char temperature was 300°C with a ramp rate of 10°C/s and a hold time of 3 s. For atomisation, the temperature was set to 1900°C with ramp and hold times of 1 and 2 s, respectively. Precision in terms of the relative standard deviation was typically <5% for triplicate injections. Three sample replicates were taken from pre-filtered (Whatman 0.45 μm cellulose nitrate membrane filters) sub-samples.

25.3.4 Control of algal growth

Visual algal cover estimations were undertaken by using a 0.2 × 0.2 m reference metal grid located on top of the attenuation pond during the experiment that could also be used as a permanent safety barrier. Microorganisms including algae such as *Microspora* spp. (dominates 'blanket weed') were counted using a Sedgewick-Rafter Cell S50 (counting chamber: 1 × 50 × 20 mm) and a Wang Biomedical Research Microscope 6000 (bright field and phase-contrast).

Excess algae and decomposing litter were occasionally removed from the attenuation pond and infiltration basin. Wet algae and litter were weighed after the biomass had drained for 2 min by applying maximum pressure with both hands to small portions of the organic waste. Blanket weed control with barley straw bales (commercial product called Frogmat) was also practised.

The straw bales were sub-divided into smaller bales before application. The new bales (approximately 0.2 kg straw each) were located at eight sites (approximately 1.6 kg straw in total) near the attenuation pond margins, and remained there almost fully submerged between 31 August and 31 October 2001. This procedure was repeated between 22 February and 18 May 2002.

Between 15 March and 12 August 2002, an experimental GAC filter cleaned the stormwater before it reached the silt trap. Approximately 1.4 kg GAC (Grade 207 EA, US mesh: 12 × 40) was used to test the purification potential with respect to nutrients and trace elements that might contribute to algal growth.

25.3.5 System capacity

The silt trap and attenuation pond had a combined total effective volume of approximately 1.9 m^3 during storm events. The rainwater drained into the silt trap with a maximum capacity of 0.1 m^3. Suspended solids (e.g., weathered building materials, decayed leaves, bird droppings and particles from atmospheric pollution) from the roofs settled predominantly in the silt trap.

Water from the silt trap overflowed into the attenuation pond (volume of approximately 1.7 m^3), which also served the purpose of a storage pond. If the attenuation pond was full, water flowed into the infiltration basin, which accommodated a maximum volume of approximately 1.8 m^3 during heavy storm events before it overflowed (Fig. 25.1).

25.4 Results and discussion

25.4.1 Standard design considerations

Tables 25.1–25.3 summarise the calculation procedures for the BRE, CIRIA and ATV-DVWK design guidelines (ATV-DVWK, 2002; Bettes, 1996; BRE, 1991;

Martin et al., 2000). Findings are based on the design assumption that flooding for the case study site might statistically occur only once within ten years (defined as the return period). The estimated mean IR applied for all methods was 10^{-4} m/s (Tables 25.1–25.4).

Table 25.1 Calculation of the required infiltration basin storage volume (italic number) using the Building Research Establishment guidelines (BRE, 1991)

Duration	min	10	15	30	60	120
Factor Z_1	–	0.46	0.56	0.75	1.00	1.30
Design rainfall M5-D (five-year return period)	mm	9.2	11.2	15.0	20.0	26.0
Growth factor Z_2	–	1.22	1.22	1.24	1.24	1.24
Design rainfall M10-D (ten-year return period; calculated from M5 rainfall)	mm	11.2	13.7	18.6	24.8	32.2
Inflow (ten-year return period)	m^3	0.97	1.18	1.60	2.13	2.77
Outflow drainage	m^3	0.12	0.18	0.36	0.72	1.44
Required storage (ten-year return); M10-D	m^3	0.85	1.00	1.24	*1.41*	1.33

Assumptions: ratio = 0.24 (ratio between the M5-60 storm and the M5-2D rainfall); impermeable area = 86 m^2; M5-60 = 20 mm; infiltration rate = 0.0001 m/s; perimeter = 10 m; depth = 0.4 m; surface area to 50% effective depth (excluding base) = 2 m^2.

Table 25.2 Calculation of the required infiltration basin storage height (italic number) using the Construction Industry Research and Information guidelines (Bettes, 1996)

Duration	min	10	15	30	60	120
Intensity of storm	mm/h	67.3	54.6	37.2	24.8	16.1
Maximum height of water to be stored	m	0.154	0.184	*0.213*	0.210	0.142

Assumptions: infiltration rate = 0.0001 m/s; safety factor = 2 (area to be drained <100 m^2; minor inconveniences, if the system fails); area of base = 4.95 m^2; perimeter = 10 m; impermeable area = 86 m^2; void ratio of the soakaway fill material = 1 (represents air).

Table 25.3 Calculation of the required infiltration basin storage volume (italic number) using the German Association for Water, Wastewater and Waste guidelines (ATV-DVWK, 2002)

Duration	min	10	15	30	60	120
Runoff for each design storm	l/(s × h)	186.9	151.7	103.3	68.9	44.7
Volume to be stored	m^3	0.85	0.99	1.19	*1.27*	0.96

Assumptions: impermeable area = 86 m^2; surface factor = 0.9 (considering roof angle and building material); infiltration area = 4.95 m^2; infiltration rate = 0.0001 m/s; safety factor = 1.1 (minor inconvenience, if the system fails).

Table 25.4 Summary statistics: water quality of the attenuation pond and infiltration rates (IR) for the infiltration basin (26 April 2001 to 12 August 2002)

Variable	Unit	Time	Sample number	Mean	1st summer[3] mean	Autumn[4] mean	Winter[5] mean	Spring mean[6]	2nd summer mean[7]
Air temperature	°C	PM[9]	345	12.2	16.9	8.9	6.5	13.2	17.1
Water temperature	°C	PM[9]	325	11.1	15.4	8.6	9.1	11.9	15.3
BOD_5[1]	mg/l	AM[10]	57	4.3	4.2	3.1	5.6	6.7	2.5
Suspended solids	mg/l	AM[10]	40	46.8	132.5	51.3	3.8	54.8	3.4
Total solids	mg/l	AM[10]	40	193.3	238.7	293.8	104.7	151.5	91.7
Conductivity	µS	PM[9]	281	39.8	75.1	37.5	33.0	37.7	34.2
Turbidity	NTU[8]	AM[10]	39	2.8	4.3	2.7	2.2	2.0	1.5
Dissolved oxygen	mg/l	PM[9]	319	9.3	11.5	4.8	10.1	12.4	7.3
pH	–	PM[9]	263	7.77	8.36	7.27	7.86	8.69	6.84
Algal cover	%	PM[9]	306	44	61	50	36	39	38
IR (Station 1)[2]	10^{-7} m/s	PM[9]	23	8	3	23	1	2	5
IR (Station 2)[2]	10^{-7} m/s	PM[9]	23	158	4270	95	42	36	580
IR (Station 3)[2]	10^{-7} m/s	PM[9]	24	9946	8152	4778	1286	6342	31841

[1]BOD_5 = five-day @ 20°C biochemical oxygen demand; [2]IR = infiltration rates were calculated for the soil layers located 20 cm below the bottom of the infiltration basin centre (Station 1), the middle of the slope of the infiltration basin (Station 2) and the lawn (Station 3), see also Fig. 25.1; [3]summer: 21/06–21/09/01; [4]autumn: 22/09–20/12/01; [5]winter: 21/12/01–19/03/02; [6]spring: 20/03–20/06/02; [7]summer: 21/06–12/08/02; [8]NTU = nephelometric turbidity unit; [9]PM = afternoon; [10]AM = morning.

The critical storm durations for the BRE, CIRIA and ATV-DVWK design calculations were 1.0, 0.5 and 1 h, respectively. The associated maximum infiltration basin height requirements were 28, 21 and 26 cm, respectively. Furthermore, the maximum infiltration storage volume requirements were 1.41, 1.04 and 1.27 m³, respectively.

Water level fluctuations within the attenuation pond and infiltration basin are indicated in Fig. 25.2. Maximum infiltration basin height requirements calculated according to BRE (1991), CIRIA (Bettes, 1996) and ATV-DVWK (2002) guidelines were not sufficient for the period of the experiment. Figure 25.2 indicates that the wetland system would have failed, if the recommended design depths for the infiltration basin had been applied.

Equation (25.4.1) indicates the mathematical relationship between the infiltration basin design depth D (mm) and the infiltration time T (h). The corresponding R for the function is 0.92.

$$D = -a + \ln(T) + b \tag{25.4.1}$$

where:

$a = 2.54 \times b^{0.51}$, R = 0.58 (in general)
$a = 0.59 \times b^{0.83}$, R = 0.89 (for b < 190 mm)
$a = 0.015 \times b^{1.51}$, R = 0.77 (for b ≥ 190 mm during spring, summer and autumn)
$a = 0.001 \times b^{1.94}$, R = 0.75 (for b ≥ 190 mm during winter)

Figure 25.3 indicates the mathematical relationship between the terms 'a' and 'b'. Both 'a' and 'b' are based on logarithmic trendline functions for 36 recorded storms (between 3 and 8 depth measurements each) between May 2001 and August 2002. Term 'a' is a coefficient, whereas 'b' represents the maximum depth (mm) within the infiltration basin during an individual storm. The corresponding R accounts for differences in infiltration behaviour for systems that are either shallower or deeper than approximately 185 mm. The mathematical relationships summarised in Eq. (25.4.1) and Fig. 25.3 give the design engineer the opportunity to estimate design depths related to different infiltration times and vice versa.

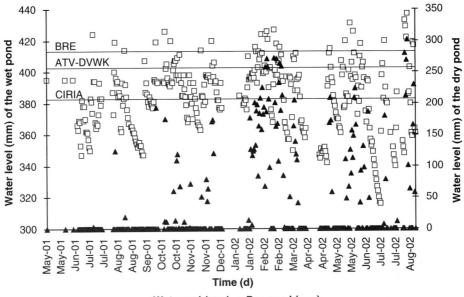

□ Wet pond (mm)   ▲ Dry pond (mm)

Fig. 25.2 Maximum daily water level fluctuations within the wet pond and dry pond between 13 May 2001 and 12 August 2002. Maximum infiltration basin height requirements calculated according to Building Research Establishment, BRE (1991), Construction Industry Research and Information Association, CIRIA (Bettes, 1996; Martin et al., 2000), and the German Association for Water, Wastewater and Waste, ATV-DVWK (ATV-DVWK, 2002), guidelines are indicated by horizontal lines.

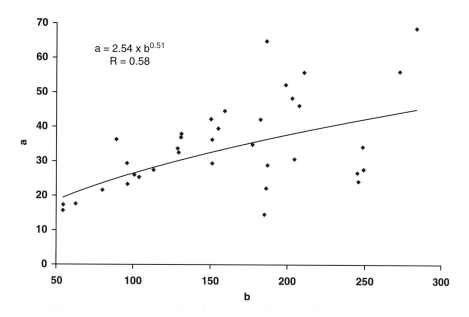

Fig. 25.3 Relationships between the coefficient 'a' and the maximum depth 'b' (mm) within the infiltration basin during an individual storm.

25.4.2 System design comparisons

The actual design for the infiltration basin (40 cm depth and 1.7 m³ volume) was acceptable when compared to BRE (1991), CIRIA (Bettes, 1996) and ATV-DVWK (2002) guidelines. Signs of system failure (e.g., flooding of the nearby lawn and structural damage) have not yet been observed. However, strict application of all test guidelines (without adding a higher safety factor than recommended) would have led to system failures during the first year of operation. This finding is in contrast to a previous investigation by Abbott and Comino-Mateos (2001), where British design guidelines did not fail. However, a perforated concrete ring soakaway to infiltrate stormwater runoff was used.

Figure 25.2 shows that all three official designs (Tables 25.1–25.3) failed at least once during the first 15 months of the ten-year design period. However, the period of study can be described as a particularly 'wet time' (2160 mm precipitation between summer 2001 and spring 2002). Moreover, the study site is situated on a semi-exposed location approximately 240 m above sea level. In comparison, the annual average rainfall for Bradford (located in a valley) is approximately 830 mm (Meteorological Office, 2002).

Considering a maximum depth of 40 cm, the infiltration basin is associated with under-utilised storage capacity. However, observations have shown that a depth of 40 cm is justified considering that 20, 10 and 1 runoff events have led to the maximum design water depths of 21, 26 and 28 cm, respectively, being exceeded (Fig. 25.2). It follows that there would have been no spare capacity for strict applications of the standard design guidelines.

25.4.3 Water quality management

The water quality of the attenuation pond (Tables 25.4 and 25.5) was acceptable for disposal (e.g., sustainable drainage) and recycling (e.g., irrigation, toilet flushing and washing of cars) according to findings presented by Butler and Davies (2000) and Ellis et al. (2002).

Figures 25.4 and 25.5 show results from the analysis of waters and digests by ICP-OES. Measured elemental concentrations were either low (boron, barium, calcium, magnesium, manganese and zinc), close to the detection limit (aluminium, copper and iron) or below the detection limit (most heavy elements were analysed but not listed). Figure 25.5 indicates that the liquid phase (rainwater) within the treatment train can actually take up soluble contaminants from the sediment; the sediment acts as a source, thereby polluting the liquid phase.

Table 25.5 Summary statistics: 24-h sampling on 15 March 2002

Variable	Unit	Total number	Mean				Standard deviation			
			00:00–05:30	06:00–11:30	12:00–17:30	18:00–23:30	00:00–05:30	06:00–11:30	12:00–17:30	18:00–23:30
Water temperature	°C	23	3.4	4.3	7.4	4.8	0.6	2.1	0.7	1.4
BOD_5[1]	mg/l	36	1.3	1.1	0.7	2.0	0.8	0.6	0.5	1.1
Nitrate-N[2]	mg/l	40	0.54	0.38	0.39	0.47	0.37	0.18	0.17	0.19
Ammonia-N	mg/l	40	0.06	0.03	0.04	0.07	0.06	0.04	0.05	0.07
Phosphate-P	mg/l	40	0.07	0.04	0.06	0.13	0.06	0.04	0.06	0.09
Total solids	mg/l	43	99.8	97.8	180.7	139.2	18.8	14.6	59.1	31.3
Conductivity	μS	24	36.0	42.0	43.1	41.3	1.2	1.5	1.7	2.0
Dissolved oxygen	mg/l	24	10.0	9.8	11.3	11.6	0.7	1.2	0.8	1.3
pH	–	24	7.54	7.57	7.89	8.01	0.39	0.51	0.24	0.58
Calcium[3,4]	mg/l	21	10.37	10.89	11.08	10.96	1.19	0.96	1.23	0.07
Magnesium[3,4]	mg/l	21	1.13	1.14	1.14	1.14	0.08	0.07	0.04	0.06
Manganese[3,4]	mg/l	21	0.006	0.006	0.005	0.010	0.002	0.002	0.002	0.009
Zinc[3,4]	mg/l	21	0.007	0.010	0.012	0.010	0.003	0.005	0.009	0.001
Zinc[3,5]	mg/l	39	0.009	0.006	0.006	0.006	0.004	0.004	0.003	0.002
Barium[3,4]	mg/l	21	1.57	1.57	1.66	1.72	0.24	0.09	0.28	0.31

[1]BOD_5 = five-day @ 20°C biochemical oxygen demand; [2]concentrations include nitrite; [3]filtered sample (0.45 μm pore size); [4]analysis with the Inductively Coupled Plasma Optical Emission Spectrometer; [5]analysis with the Electrothermal Atomic Absorption Spectrometer.

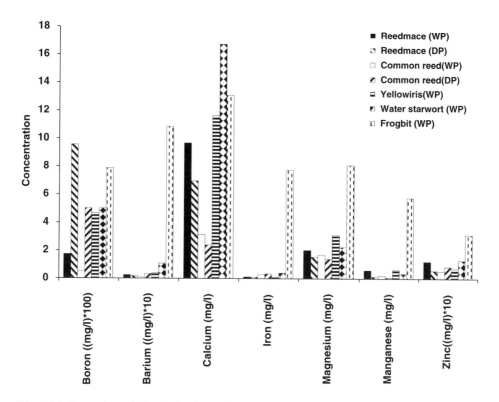

Fig. 25.4 Summary of the Inductively Coupled Plasma Optical Emission Spectrometer results for different aquatic plants (dry weights) located in the attenuation pond and infiltration basin. Concentration readings for boron, barium and zinc require division by a factor of 100, 10 and 10, respectively. WP = wet pond. DP = Dry pond.

25.4.4 Twenty-four hour water quality monitoring

Manual sampling over 24 h was conducted on 15 March 2002 (Table 25.5). Most samples were taken either in 30-min intervals or hourly. Despite the cold climate, diurnal water quality variations (Wu and Mitsch, 1998) were apparent for water temperature, BOD_5, nitrate-N (including nitrite-N), ammonia-N, conductivity, pH, manganese, zinc and barium.

Of the elements detected by ICP-OES, zinc was the only one that had a high overall standard deviation (approximately 0.0060 mg/l, for triplicate analyses of each sample) as measurements were made close to the detection limit. However, analysing the same samples using ETAAS reduced the overall standard deviation to 0.0036 mg/l (three replicates for each sample) as this technique is more sensitive for the determination of zinc. There were no significant fluctuations of zinc during the day.

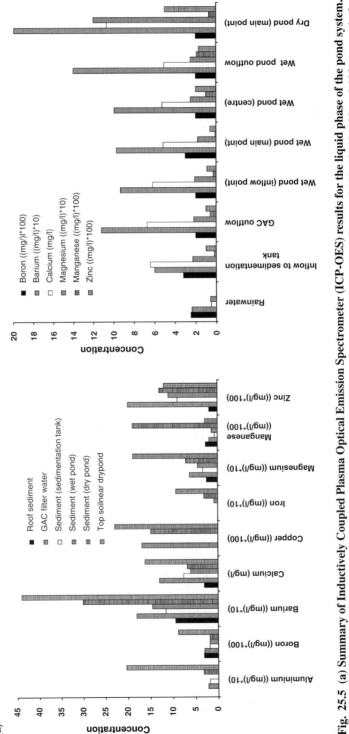

Fig. 25.5 (a) Summary of Inductively Coupled Plasma Optical Emission Spectrometer (ICP-OES) results for the sediment phase of the pond system. Concentration readings for boron, barium, magnesium, manganese and zinc require division by a factor of 10, 100, 10, 100 and 100, respectively, and (b) summary of ICP-OES results for the liquid phase of the pond system. Concentration readings for aluminium, boron, barium, copper, iron, magnesium, manganese and zinc require division by a factor of 10, 100, 10, 100, 10, 10, 100 and 100, respectively. GAC = granular activated carbon.

25.4.5 Aquatic plant management

To prevent algal washout into the infiltration basin during storm events, it is necessary to control the rapid growth of algae with barley straw and by mechanical removal (P. Taylor, Director, Frogmat International Holdings, personal communication). Algal washout allows decaying algae to block fine pores, thereby decreasing the infiltration capacity of the basin (see also Table 25.4).

In addition to biological (grazing zooplankton and tadpoles) and physical (removal by hand) algal control, biochemical control was also implemented (with barley straw extracts). The presence of small barley straw bales is likely to reduce algal growth by releasing a 'cocktail' of phytotoxic chemicals (Everall and Lees, 1997).

The wet weights of litter removed during summer 2001, autumn 2001, winter 2001/02 and spring 2002 were 1.1, 4.6, 1.0 and 6.2 kg, respectively. The number of occasions on which plant harvesting was carried out during these seasons were 6, 10, 3 and 14, respectively.

Litter production depends on the season (Scholz and Xu, 2002; Wu and Mitsch, 1998). Fresh biomass is usually produced in spring and summer, and requires harvesting in late autumn. Furthermore, algae (predominantly *Microspora* spp.) require frequent removal in spring. In contrast to summer and winter, the maintenance was more efficient during autumn and spring, because less labour time (i.e. work input expressed as time) was required, and more litter and algae were removed during these maintenance sessions.

Figure 25.4 indicates the elemental content of various aquatic plants. It can be seen that macrophytes located in the attenuation pond contain higher concentrations of calcium, magnesium and manganese in comparison to macrophytes located in the infiltration basin. Furthermore, frogbit was associated with high elemental concentrations in general. This was also the case for barium, iron, magnesium, manganese and zinc.

All assessed aquatic plants were associated with detectable concentrations of magnesium, manganese and zinc. Measured concentrations of these elements for aquatic plants were relatively higher than for the liquid phase (pond water), but comparable to associated sediment samples (Figs. 25.4 and 25.5). However, a direct comparison between different loadings and concentrations (mg/kg and mg/l, respectively) is difficult. Nevertheless, plant harvesting reduced trace nutrients and elements (in addition to BOD_5, SS, etc.) in the attenuation pond as indicated previously by Wu and Mitsch (1998).

25.5 Conclusions

The case study described the successful design, operation and maintenance of a novel stormwater pond system during the first 15 months of operation. However,

other filter media with high adsorption capacities such as activated carbon (Scholz and Martin, 1998a; Scholz et al., 2002) and oxide-coated sand (Sansalone, 1999) has been discussed elsewhere.

Filters 11 and 12 are more complex in their design and operation (Table 26.1). The top water layer of both filters is aerated (with air supplied by air pumps) to enhance oxidation (minimizing zones of reducing conditions) and nitrification (Cheng et al., 2002; Green et al., 1998; Obarska-Pempkowiak and Klimkowska 1999). Filter 12 receives about 153% of Filter 11's mean annual inflow volume and load (Table 26.1). The hydraulic regime of Filter 12 differs from that of Filters 1–11 in order to identify the best filtration performance. A higher hydraulic load should result in greater stress on *P. australis* and biomass.

26.3.3 Environmental conditions and operation

The filtration system was designed to operate in batch flow mode to reduce pumping and computer control costs. All filters were periodically inundated with pre-treated inflow gully pot liquor, and drained to encourage air penetration through the aggregates (Cooper et al., 1996; Gervin and Brix, 2001; Scholz and Xu, 2002).

Since 22 September, 2003, the pH value of the inflow has been artificially raised by addition of sodium hydroxide (NaOH) to the sieved gully pot liquor to prevent a breakthrough of metals (see below and Tchobanoglous et al., 2003). It follows that the inflow pH was therefore increased from a mean pH 6.7 to a mean pH 8.1 (Table 26.3).

26.3.4 Metal nitrates

Copper and nickel were selected as additional heavy metals for investigation, because they are commonly occurring contaminants from road vehicles and are not easily bio-available (Cooper et al., 1996; Kadlec and Knight, 1996; Scholz and Xu, 2002). It follows that these metals are likely to accumulate within the sediment and debris of constructed wetlands. As the build-up continues, metal toxicity increases as does the risk of severe pollution due to leaching (Scholz et al., 2002).

Some heavy metals do accumulate easily in constructed wetlands, but may be released, if environmental conditions change; e.g., road gritting (containing salt) in winter. Such transformation processes are not well understood (Norrstrom and Jacks, 1998).

Copper nitrate and nickel nitrate were added to the inflow water of Filter 2 and Filters 7–12 to give total concentrations of dissolved copper and nickel of approximately 1 mg/l for each metal, which is comparable to the figures reported for urban water heavily contaminated with heavy metals and mine wastewater (Cooper et al., 1996; Kadlec and Knight, 1996; Mungur et al., 1997; Scholz and Xu, 2002).

Concerning the dosed inflow water, the background concentration for nitrate-nitrogen (including nitrite-nitrogen) was only approximately 0.497 mg/l. Therefore, introduced nitrate-nitrogen contributed to 65% (or approximately 0.917 mg/l) of the overall nitrate-nitrogen (including nitrite-nitrogen) load.

The filter volumes available for influent water differ among the filters due to different filter media compositions (Table 26.2) and hydraulic regimes (Table 26.1). The mean annual total loading rates for the contaminated filters were therefore between 96 and 187 mg for each metal.

26.3.5 Metal determinations

Metal concentrations were determined in the raw gully pot liquor, sieved (pore size of 0.25 mm) gully pot liquor (partially used as actual inflow water for some filters), contaminated (added metal nitrates) sieved gully pot liquor (partially used as actual inflow water for some filters) and the outflow waters from the experimental rig (Table 26.1). Raw gully pot liquor was sieved to simulate preliminary and primary treated stormwater. This procedure is in line with common practice in the wastewater industry (Cooper et al., 1996; Tchobanoglous et al., 2003).

A Varian Spectr AA 400 Atomic Absorption Spectrometer with a GTA-96 graphite furnace tube atomizer was used for the standard analysis of nickel and copper. Notched GTA partition tubes (coated) were applied, and the carrier gas was argon.

26.3.6 BOD, nutrient and other determinations

The BOD was determined in all water samples with the OxiTop IS 12-6 system, a manometric measurement device, supplied by the Wissenschaftlich-Technische Werkstätten (WTW), Weilheim, Germany. Nitrification was suppressed by adding 0.05 ml of 5 g/l N-Allylthiourea (WTW Chemical Solution No. NTH 600) solution per 50 ml of sample water.

Nitrate was reduced to nitrite by cadmium and determined as an azo dye at 540 nm (using a Perstorp Analytical EnviroFlow 3000 flow injection analyzer) following diazotisation with sulfanilamide and subsequent coupling with N-1-naphthylethylendiamine dihydrocloride (Allen, 1974).

Ammonia-nitrate and ortho-phosphate-phosphorus were determined by automated colorimetry in all water samples from reaction with hypochlorite and salicylate ions in solution in the presence of sodium nitrosopentacyanoferrate (nitroprusside), and reaction with acidic molybdate to form a phosphomolybdenum blue complex, respectively (Allen, 1974). The colored complexes formed were measured spectrometrically at 655 and 882 nm, respectively, using a

Bran and Luebbe autoanalyzer (Model AAIII). All other analytical procedures were performed according to the American Standard Methods (Clesceri et al., 1998). Composite water samples were analyzed on Mondays, Wednesdays and Fridays.

26.4 Experimental results and discussion

26.4.1 Inflow water quality

Table 26.3 summarises the water quality of the inflow and those filters artificially contaminated with heavy metals after the first and second year of operation. The pH of the inflow was artificially raised to assess its influence on the treatment performance and particularly on the potential breakthrough of heavy metals during the second winter.

Raw gully pot liquor was sieved (pore size of 2.5 mm) to simulate preliminary treatment (Tchobanoglous et al., 2003). Sieving resulted in a mean annual reduction of BOD and SS by approximately 12 and 22%, respectively.

The inflow data set was divided into two sub-sets (winter and summer) to assess the effect of seasonal variations (e.g., temperature) and road management (e.g., road gritting and salting) on the water quality. Most variables including BOD (except for the first year of operation), SS, TS, turbidity and conductivity are high in winter compared to summer (Table 26.3).

26.4.2 Comparison of annual outflow water qualities

The overall filtration performance figures are summarised in Tables 26.4 and 26.5. Figure 26.2 shows selected inflow and selected outflow concentrations for nickel and copper. Concerning BOD removal, the performances of all filters (except for Filters 1 and 2; extended storage) improved greatly over time. The reductions in BOD were also satisfactory for most filters, if compared to minimum American and European standards (<20 mg/l) for the secondary treatment of effluent (Tables 26.3–26.5).

Furthermore, the artificial increase of pH after the first year of operation had no apparent influence on the treatment performance of BOD. There is no obvious difference in performance between Filters 8 and 11, indicating that aeration did not contribute significantly to the removal of BOD (Tables 26.3 and 26.4).

Negative reduction rates for TS and conductivity were predominantly caused by road salting in late autumn and winter. Any conventional filter system including constructed wetlands is unable to retain salts in high concentrations. Therefore, salts can not be retained after a certain loading threshold, which is associated with a lag period, is exceeded. The lag period is predominantly a function of the buffering capacity of the biomass and the batch-flow operational mode (see above).

Table 26.3 Primary treated gully pot effluent: water quality variables after contamination with hydrated copper nitrate and hydrated nickel nitrate

			22/09/02–21/09/03				
Variable	Unit	Number of samples	Mean	SD[a]	Mean (winter)	Mean (summer)	
Nickel (dissolved)	mg/l	57	1.06	0.143	1.08	1.09	
Copper (dissolved)	mg/l	58	1.03	0.036	1.04	1.02	
BOD[b]	mg/l	58	61.1	49.29	43.8	86.9	
Nitrate-nitrogen	mg/l	63	1.45	1.008	1.43	1.12	
Ammonia-nitrogen	mg/l	63	1.65	2.058	1.72	2.11	
Ortho-phosphate-phosphorus	mg/l	62	0.06	0.149	0.05	0.03	
Suspended solids	mg/l	70	335.7	377.75	743.7	160.7	
Total solids	mg/l	66	2995.5	6793.27	9403.9	376.5	
Turbidity	NTU	71	311.7	479.65	690.6	162.1	
Dissolved oxygen	mg/l	68	4.70	2.493	5.70	3.16	
pH	–	71	6.69	0.411	6.89	6.72	
Redox potential	MV	62	142.5	112.72	165.5	2.7	
Conductivity	μS	71	5139.1	11182.22	15311.6	501.0	
Temperature (air)	°C	70	12.3	5.96	7.2	17.9	
Temperature (gully pot)	°C	69	10.7	5.78	4.8	18.0	

			22/09/03–21/09/04 (artificial increase of pH after 21/09/03)				
Variable	Unit	Number of samples	Mean	SD[a]	Mean (winter)	Mean (summer)	
Nickel (dissolved)	mg/l	68	1.02	0.042	1.01	1.02	
Copper (dissolved)	mg/l	66	1.02	0.018	1.01	1.02	
BOD[b]	mg/l	73	89.2	55.5	89.7	66.6	
Nitrate-nitrogen	mg/l	72	1.38	1.220	1.19	1.14	
Ammonia-nitrogen	mg/l	72	1.45	1.934	1.89	1.35	
Ortho-phosphate-phosphorus	mg/l	72	0.10	0.136	0.07	0.16	
Suspended solids	mg/l	75	853.9	1420.85	1955.2	366.7	
Total solids	mg/l	71	2141.8	3219.84	5296.4	794.7	
Turbidity	NTU	78	274.5	358.57	546.2	143.6	
Dissolved oxygen	mg/l	78	3.07	1.49	3.41	3.22	
pH	–	78	8.07	1.082	8.25	8.73	
Redox potential	Mv	78	44.4	93.44	31.8	64.1	
Conductivity	μS	78	2227.2	4620.82	6191.7	403.6	
Temperature (air)	°C	155	13.7	6.46	7.1	20.6	
Temperature (gully pot)	°C	75	12.0	5.65	6.0	18.9	

[a] standard deviation.

[b] five-day @ 20°C N-Allylthiourea biochemical oxygen demand.

Abbreviation: na = not available.

It follows that after an initial positive removal period, the removal efficiencies are becoming negative (Norrstrom and Jacks, 1998). Furthermore, the dissolved solids fraction increases as microbial biomass mineralizes the organic contaminants. Conductivity correlates well with dissolved solids that contribute to a large proportion of the TS mass (Cooper et al., 1996; Scholz et al., 2002).

Table 26.4 Mean and standard deviation of outflow water quality variables

		Mean (22/09/02–21/09/03)					
Variable	Unit	Filter 1	Filter 2	Filter 3	Filter 4	Filter 5	Filter 6
BOD[a]	mg/l	37.2	43.4	16.4	30.7	23.3	33.5
Suspended solids	mg/l	174.6	189.0	120.8	130.4	132.4	127.3
Total solids	mg/l	2772.8	3602.5	2938.7	3266.2	2949.1	3990.4
Turbidity	NTU	79.8	89.0	15.0	39.0	35.4	25.6
Conductivity	μS	5148.5	6827.7	5920.8	5392.6	5809.9	5797.7
Variable	Unit	Filter 7	Filter 8	Filter 9	Filter 10	Filter 11	Filter 12
BOD[a]	mg/l	12.7	22.0	21.8	37.6	20.1	19.8
Suspended solids	mg/l	96.7	92.5	85.8	79.1	73.2	163.5
Total solids	mg/l	2375.6	2118.4	1717.1	1808.8	1779.0	2941.7
Turbidity	NTU	11.2	20.6	27.3	34.3	17.5	43.5
Conductivity	μS	4912.2	4346.8	3808.8	3667.8	3528.5	5941.4

		Mean (22/09/03–21/09/0404; artificial increase of pH after 21/09/03)					
Variable	Unit	Filter 1	Filter 2	Filter 3	Filter 4	Filter 5	Filter 6
BOD[a]	mg/l	30.2	30.4	3.1	2.3	4.4	4.4
Suspended solids	mg/l	434.6	907.5	76.6	65.8	130.4	82.7
Total solids	mg/l	1690.3	1931.8	1379.6	1398.9	1372.9	1430.7
Turbidity	NTU	118.6	117.8	6.0	4.8	8.6	6.0
Conductivity	μS	2268.2	2356.0	2339.9	2260.3	2220.0	2507.6
Variable	Unit	Filter 7	Filter 8	Filter 9	Filter 10	Filter 11	Filter 12
BOD[a]	mg/l	2.3	2.7	2.8	3.5	3.6	7.6
Suspended solids	mg/l	147.3	78.1	83.4	71.7	89.9	102.8
Total solids	mg/l	1390.3	1497.3	1509.4	1804.7	1578.5	1647.4
Turbidity	NTU	9.6	6.9	10.2	4.9	7.4	20.1
Conductivity	μS	2199.2	2484.2	2450.6	2459.6	2534.5	2495.3

		Standard deviation (22/09/02–21/09/03)					
Variable	Unit	Filter 1	Filter 2	Filter 3	Filter 4	Filter 5	Filter 6
BOD[a]	mg/l	39.89	41.95	13.09	31.24	22.74	30.33
Suspended solids	mg/l	365.53	440.59	265.83	252.75	249.34	273.44
Total solids	mg/l	7273.37	10821.20	6886.58	6585.16	6216.36	8318.42
Turbidity	NTU	96.15	73.83	24.21	61.19	46.64	28.45
Conductivity	μS	13473.42	19920.19	13442.84	11789.78	12286.69	12860.16
Variable	Unit	Filter 7	Filter 8	Filter 9	Filter 10	Filter 11	Filter 12
BOD[a]	mg/l	13.88	25.03	20.08	35.76	14.04	17.63
Suspended solids	mg/l	225.06	197.81	168.08	166.03	144.40	307.73
Total solids	mg/l	5948.07	4720.47	3580.56	3648.16	3659.61	6450.13
Turbidity	NTU	18.95	30.18	42.51	54.91	18.07	86.23
Conductivity	μS	11672.86	9431.98	6319.37	6683.83	7077.11	11479.77

Table 26.4—(cont'd)

Variable	Unit	Filter 1	Filter 2	Filter 3	Filter 4	Filter 5	Filter 6
		\multicolumn{6}{c}{Standard deviation (22/09/03–21/09/0404; artificial increase of pH after 21/09/03)}					
BOD[a]	mg/l	21.24	37.06	7.23	3.11	8.11	6.66
Suspended solids	mg/l	449.80	3580.34	96.29	79.97	263.65	107.28
Total solids	mg/l	1507.80	2417.22	1722.92	1601.87	1445.83	1714.86
Turbidity	NTU	90.33	258.22	6.89	4.44	8.38	4.48
Conductivity	μS	2945.92	3000.52	2839.88	2565.04	2476.56	2857.82

Variable	Unit	Filter 7	Filter 8	Filter 9	Filter 10	Filter 11	Filter 12
BOD[a]	mg/l	3.86	3.83	4.00	4.49	4.24	7.83
Suspended solids	mg/l	228.84	82.75	96.24	86.77	133.64	138.81
Total solids	mg/l	1623.29	1538.09	1716.03	2553.49	1905.64	2426.29
Turbidity	NTU	18.79	7.76	19.59	4.24	10.94	37.18
Conductivity	μS	2611.49	2572.43	2840.91	2741.12	3069.47	3638.43

[a] five-day @ 20°C N-Allylthiourea biochemical oxygen demand.

In contrast to previous researchers, who reported the worst seasonal performance for BOD removal during winter (Karathanasis et al., 2003), all filters with the exception of Filters 1 and 2 showed high BOD removal figures ($>94\%$) even in the second winter. This suggests that soil microbes still have the capacity to effectively decompose organic matter in winter.

Concerning other variables, reduction rates for SS increased also in the second year although outflow concentrations frequently exceeded the threshold of 30 mg/l throughout the year except for summer. Turbidity values of the outflow decreased greatly over time. Despite the artificial increase of pH in the inflow, the pH of the outflow was approximately neutral and comparable to the first year of operation. Moreover, the pH of the outflow was relatively stable in the second year (standard deviation of approximately 0.18).

26.4.3 Heavy metal removal

Heavy metal removal efficiencies improved during the second year of operation (Fig. 26.2). However, the reduction in metals was not sufficient to comply with American standards for secondary wastewater treatment (Clesceri et al., 1998). Dissolved nickel and dissolved copper concentrations should not exceed 0.0071 and 0.0049 mg/l, respectively (Tchobanoglous et al., 2003).

The decomposition of aquatic plants after autumn, reducing soil conditions, road gritting and salting during periods of low temperatures, and acid rain contribute to increases of metal concentrations in the outflow (Norrstrom and Jacks, 1998; Sasaki et al., 2003). For example, high levels of conductivity were recorded

Table 26.5 Removal (%) per wetland filter[a] of outflow variables

22/09/02–21/09/03

Variable	Filter 1			Filter 2			Filter 3			Filter 4			Filter 5			Filter 6		
	O[b]	W[c]	S[d]	O[b]	W[c]	S[d]	O[b]	W[c]	S[d]	O[b]	W[c]	S[d]	O[b]	W[c]	S[d]	O[b]	W[c]	S[d]
BOD[e]	43	13	82	32	3	80	76	59	85	51	40	84	65	49	88	49	44	86
SS[f]	52	40	81	44	25	78	67	53	92	65	54	86	64	54	92	66	52	88
TS[g]	9	N	N	N	N	21	4	N	6	N	1	N	3	N	N	N	N	N
Turb[h]	78	86	65	71	86	51	96	98	95	90	94	94	90	96	91	93	95	93
Cond[i]	2	N	N	N	N	2	N	N	N	N	N	N	N	N	N	N	N	N

Variable	Filter 7			Filter 8			Filter 9			Filter 10			Filter 11			Filter 12		
	O[b]	W[c]	S[d]	O[b]	W[c]	S[d]	O[b]	W[c]	S[d]	O[b]	W[c]	S[d]	O[b]	W[c]	S[d]	O[b]	W[c]	S[d]
BOD[e]	79	53	94	66	32	92	67	51	93	43	2	89	69	68	89	69	65	84
SS[f]	72	60	98	73	62	91	75	71	95	77	70	92	78	71	88	51	35	86
TS[g]	20	15	25	29	32	N	42	48	19	39	45	2	40	43	N	2	N	28
Turb[h]	96	97	99	93	95	96	91	94	98	89	90	93	94	96	94	86	85	92
Cond[i]	4	N	N	15	11	N	26	36	N	28	32	N	31	29	N	N	N	N

22/09/03–21/09/0404 (artificial increase of pH after 21/09/03)

Variable	Filter 1			Filter 2			Filter 3			Filter 4			Filter 5			Filter 6		
	O[b]	W[c]	S[d]	O[b]	W[c]	S[d]	O[b]	W[c]	S[d]	O[b]	W[c]	S[d]	O[b]	W[c]	S[d]	O[b]	W[c]	S[d]
BOD[e]	66	69	76	63	44	71	97	100	96	97	99	94	95	99	96	95	99	93
SS[f]	49	74	52	15	N	76	91	91	98	92	92	98	85	91	97	90	91	97
TS[g]	25	36	37	15	23	56	38	40	65	38	42	58	39	50	64	36	38	49
Turb[h]	57	77	48	61	68	62	98	99	95	98	99	98	97	97	96	98	99	98
Cond[i]	N	7	18	N	8	N	N	12	N	N	18	N	0	22	N	N	9	N

Variable	Filter 7			Filter 8			Filter 9			Filter 10			Filter 11			Filter 12		
	O[b]	W[c]	S[d]	O[b]	W[c]	S[d]	O[b]	W[c]	S[d]	O[b]	W[c]	S[d]	O[b]	W[c]	S[d]	O[b]	W[c]	S[d]
BOD[e]	97	99	95	97	99	91	97	99	96	96	99	90	96	98	90	91	94	88
SS[f]	81	89	99	90	91	96	91	90	99	91	91	98	89	87	97	87	87	98
TS[g]	39	42	72	34	46	50	34	43	71	21	22	53	31	36	56	28	24	34
Turb[h]	97	98	99	98	98	98	97	99	98	98	99	99	98	97	99	93	93	97
Cond[i]	3	16	N	N	17	N	N	16	N	N	15	N	N	9	N	N	2	N

[a] Change (%) $= \frac{(\text{in} - \text{out}) \times 100\,(\%)}{\text{in}}$, where in = inflow and out = outflow.

[b] overall mean.

[c] mean of the winter.

[d] mean of the summer.

[e] five-day @ 20°S N-Allylthiourea biochemical oxygen demand (mg/l).

[f] suspended solids (mg/l).

[g] total solids (mg/l).

[h] turbidity (NTU).

[i] conductivity (μS).

In italics: BOD >20 mg/l and SS >30 mg/l (outflow concentrations exceeding thresholds).

Abbreviation: N = negative removal (i.e. more output than input).

in the filter inflow and outflows (Tables 26.3–26.5), and the breakthrough of dissolved nickel was observed during the first winter (Fig. 26.2a).

Concerning the effect of retention time on the treatment efficiency of metals, the heavy metal outflow concentrations of Filter 12 (higher loading rate) were slightly higher than the corresponding concentrations for the other filters. According to previous studies (Kadlec and Knight, 1996; Wood and Shelley, 1999), metal removal efficiencies for wetlands are highly correlated with influent concentrations and mass loading rates. Moreover, it was suggested that the formation of metal sulfides was favored in wetlands with long retention times. This may lead to more sustainable management of constructed treatment wetlands.

26.4.4 Link between pH and treatment of metals

After the increase of the inflow pH, mean reduction efficiencies for nickel increased during the second winter compared to the first winter; e.g., 90 and 65%, respectively, for Filter 7 (Fig. 26.2a). Moreover, an obvious breakthrough of nickel was not observed during the second winter despite of the presence of high salt concentrations in the inflow. This is likely to be due to the artificial increase of pH.

A high pH facilitates nickel precipitation. For example, nickel hydroxide ($Ni(OH)_2$) may precipitate at pH 9.1, if the corresponding metal concentration is 1 mg/l (Tchobanoglous et al., 2003), which is similar to the inflow concentrations of spiked filters. Moreover, dissolved copper did not break through any constructed wetland filter throughout the study (Fig. 26.2b).

All filters acted as pH buffers after pH increase, and pH levels were subsequently reduced. It can be assumed that this buffering capacity is greatly enhanced by the presence of active biomass rather than macrophytes (Kadlec and Knight, 1996; Sasaki et al., 2003). However, the outflow pH values for the planted filters recorded were slightly lower than those for the unplanted filters. For example, the overall mean pH value for Filter 7 (unplanted filter) is 7.31, and the corresponding value for Filter 8 (planted filter) is 6.98 during the second year of operation.

26.4.5 Analysis of variance and modeling

An analysis of variance has shown that all filters containing aggregates are relatively similar to each other with respect to most of their outflow variables. It follows that some filters could be considered as replicates (e.g., Filter 7 is a replicate of Filter 3) despite the differences in filter set-up (Tables 26.1 and 26.2). The P values of the pairs for Filter 3 and 7 are 0.20, 0.95 and 0.98 for BOD, SS and turbidity, respectively. Pairs of data associated with P ≥ 0.05 can be regarded as statistically similar.

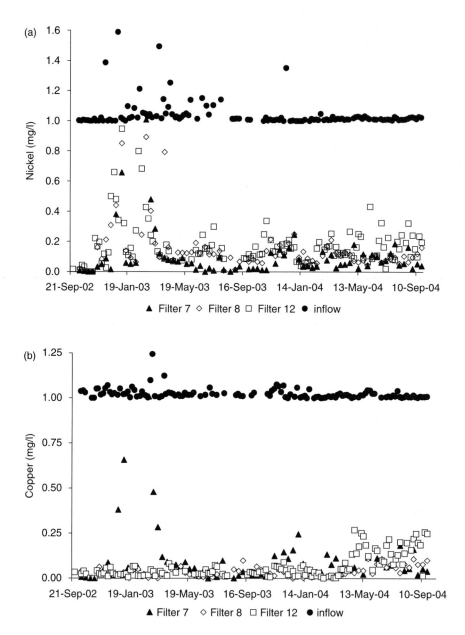

Fig. 26.2 Daily (a) nickel and (b) copper concentrations (mg/l) in the inflow and outflows of Filters 7 (unplanted), 8 (planted) and 12 (planted and high loading).

The absence of filter replicates is not a serious issue considering that the filter performances are similar (see above), and that the standard deviations (Table 26.4) of most outflow variables are comparable with large-scale systems (Cooper et al., 1996; Kadlec and Knight, 1996; Karathanasis et al., 2003; Vymazal, 2002). Moreover, a more detailed experimental study with three replicate filters, for example, would not be justified in terms of costs and potential scientific benefit.

26.5 Conclusions and further work

Despite of the highly variable water quality of road runoff, the novel filters showed great treatment performances particularly with respect to the BOD reduction in a cold climate. Removal efficiencies for SS in particular, improved over time, and dissolved copper was removed satisfactorily in comparison to values obtained from the literature. However, a breakthrough of dissolved nickel during the first winter of the first year of operation was observed. After creating an artificially high inflow pH of approximately eight after one year of operation, nickel was successfully treated despite vulnerability to leaching when exposed to a high salt concentration during the second winter.

A high pH was apparently also linked to high removal efficiencies. The elevated pH had no apparent negative effect on the biomass including macrophytes. Moreover, filters showed a great pH buffering capacity. Findings indicate that conventional pH adjustment can be successfully applied to constructed wetland systems for stormwater treatment.

The presence of Filtralite (adsorption filter media) and *P. australis* (Cav.) Trin. Ex Steud., common reed (macrophyte), did not result in an obvious reduction of metal concentrations in outflow waters. Operational conditions such as inflow pH and retention time were more important for the heavy metal treatment.

Furthermore, case-based reasoning of the experimental data set has been applied by Lee et al. (2005b). The purpose of this follow-up paper is to show how case-based reasoning and other advanced modeling techniques (see also Chapter 28) can be applied for water quality control purposes.

Chapter 27

Combined wetland and below-ground detention systems

27.1 Experimental case study summary

The aim of this study was to assess the treatment efficiencies of experimental below-ground stormwater detention (extended storage) systems receiving concentrated runoff that has been pre-treated by filtration with different aggregates as part of artificial wetlands. Randomly collected gully pot liquor was used instead of road runoff (see also Chapter 3). To test for a 'worst case scenario', the experimental system received higher volumes and pollutant concentrations in comparison to large-scale below-ground detention systems under real (frequently longer but very diluted) runoff events.

Gravel, sand, Ecosoil, block paving and turf were tested in terms of their influence on the water quality. Concentrations of 5-day @ 20°C ATU biochemical oxygen demand (BOD) in contrast to suspended solids (SS) were frequently reduced to below international secondary wastewater treatment standards. The denitrification process was not completed. This resulted in higher outflow than inflow nitrate-nitrogen concentrations.

An analysis of variance (ANOVA) indicated that some systems were similar in terms of most of their treatment performance variables including BOD and SS. It follows that there is no need to use additional aggregates with high adsorption capacities in the primary treatment stage from the water quality point of view (e.g., Ecosoil).

27.2 Introduction

27.2.1 Sustainable drainage systems

The acronym for Sustainable Urban Drainage System (British phrase) is SUDS, which is also known as Best Management Practice (American phrase). A singular or series of management structures and associated processes designed to drain surface runoff in a sustainable approach to predominantly alleviate capacities in existing conventional drainage systems in an urban environment is defined as SUDS (Butler and Davies, 2000; CIRIA, 2000; SEPA, 1999).

New developments proposed for Brownfield sites or the periphery of urban developments may be unable to obtain planning permission, if existing local sewers have no spare capacity for stormwater drainage, and if the stormwater discharge from the proposed site cannot be controlled. In the absence of suitable watercourses that can accommodate direct stormwater discharges, alternative technologies such as 'at source' stormwater storage and detention systems are required (Butler and Davies, 2000).

Maintenance of all public SUDS structures above-ground is usually the responsibility of the local authority (The Stationery Office, 1998). Above-ground SUDS structures are defined as swales, ponds, basins and any other ground depression structures. In contrast, the maintenance of below-ground SUDS structures is usually the responsibility of the regional water authority. Below-ground SUDS structures include culverts, infiltration trenches, filter strips and below-ground detention systems (Butler and Davies, 2000; CIRIA, 2000; Nuttall et al., 1997).

Stormwater runoff is usually collected in gully pots that can be viewed as simple physical, chemical and biological reactors. They are particularly effective in retaining SS (Bulc and Slak, 2003). Currently, gully pot liquor is extracted once or twice per annum from road drains and transported (often over long distances) for disposal at sewage treatment works (Butler et al., 1995; Memon and Butler, 2002). A more sustainable solution would be to treat the entire road or car park runoff locally in potentially sustainable stormwater detention systems such as below-ground storage systems and stormwater ponds (Guo, 2001) reducing transport and treatment costs. Furthermore, runoff treated with stormwater detention systems can be recycled for irrigation purposes.

Below-ground stormwater storage and detention systems are defined as a sub-surface structure designed to accumulate surface runoff, and where water is released from, as may be required to increase the flow hydrograph. The structure may contain aggregates with a high void ratio or empty plastic cells, and act also as a water recycler or infiltration device (Butler and Parkinson, 1997).

A below-ground stormwater detention system comprises a number of components forming a structure that is designed to reduce stormwater flow. The system captures surface water through infiltration and other methods. The filtered stormwater is usually stored below-ground in a tank. The water is often cleaned and filtered before it is infiltrated or discharged to the sewer or watercourse via a discharge control valve. The system benefits include runoff reduction of minor storms, groundwater recharge and pollution reduction. This detention system is predominantly applied in new developments.

27.2.2 Project purpose

The purpose of the study is to advance knowledge and understanding by formulating design guidelines for vertical-flow stormwater detention systems treating road runoff predominantly by extended storage in a cold climate such as the South-east of Scotland. The objectives are to assess the following:

- The function of turf (absent versus present) and different aggregates such as Ecosoil as components of a primary treatment filtration stage located above the below-ground detention systems; and
- The overall passive treatment performance of vertical-flow stormwater detention systems.

27.3 Materials and methods

27.3.1 System design and operation

Five stormwater detention systems (Matrix Geo-Cell provided by Atlantis Water Management Ltd.) were located outdoors at The King's Buildings campus (The University of Edinburgh, Scotland) to assess the system performance during a relatively cold year (20/03/04–19/03/05). Inflow water, polluted by road runoff, was collected by manual abstraction with a 2-l beaker from randomly selected gully pots on the campus and the nearby main roads.

Five stormwater detention systems based on plastic cells (boxes with large void space) were used. Each system had the following dimensions: height = 85 cm, length = 68 cm and width = 41 cm. Two plastic cells on top of each other made up one experimental detention system. The bottom cell (almost 50% full at any time) was used for water storage only. The top cell contained the aggregates. Different packing order arrangements of aggregates and plant roots were used in the systems (Tables 27.1 and 27.2) to test the impacts of gravel, sand, Ecosoil, block paving and turf on the water treatment performance.

The filtration system was designed to operate in vertical-flow batch mode. Manual flow control was practised. Gully pot liquor compares well with concentrated road runoff (by a factor of at least 30 depending on spacings between gully pots), and was used in the experiment as a 'worst case scenario' liquid replacing road runoff. All detention systems (Tables 27.1 and 27.2) were approximately twice per week watered with 10-l gully pot liquor as slow as possible, and drained by gravity afterwards to encourage air penetration through the soils (Cooper et al., 1996; Gervin and Brix, 2001). The relative quantity of gully pot liquor used per system was approximately $3.6 \times$ the mean annual rainfall volume to simulate again a 'worst case scenario'. The hydraulic residence times were in the order of 1 h.

**Table 27.1 Systematic and stratified experimental set-up of gravel-filled
stormwater detention system content and operation**

System	Planted	Additional media type	Natural aeration restricted
1	No	–	No
2	No	Sand	No
3	No	Sand and Ecosoil	No
4	No	Sand, Ecosoil and block paving	Yes (due to block paving)
5	Yes	Sand, Ecosoil and turf	No

**Table 27.2 Packing order of the stormwater detention systems (S) called Matrix Geo-Cells
(provided by Atlantis Water Management Ltd.)**

Height (mm)	S1	S2	S3	S4	S5
861–930 (top)	Air	Air	Air	Paving and	Air
791–860	Air	Air	Air	6 mm gravel	Turf
751–790	Air	Air	Sand and Ecosoil	Sand and Ecosoil	Sand and Ecosoil
711–750	Air	Sand	Sand and Ecosoil	Sand and Ecosoil	Sand and Ecosoil
661–710	6 mm gravel	6 mm gravel	6 mm gravel	6 mm gravel	6 mm gravel
451–660	20 mm gravel	20 mm gravel	20 mm gravel	20 mm gravel	20 mm gravel
437–450	Sand	Sand	Sand	Sand	Sand
431–436	Geotextile	Geotextile	Geotextile	Geotextile	Geotextile
201–430	Air	Air	Air	Air	Air
0–200 (bottom)	Water	Water	Water	Water	Water

Biodegradation was enhanced by encouraging natural ventilation of the
aggregates from the top via the natural air, and from the bottom via the air pockets
above the stored water and between the aggregates. Considering industrial-
scale systems, vertical ventilation pipes should be installed to encourage passive
ventilation as well.

27.3.2 Analytical methods

The BOD was determined in the inflow and outflow water samples with the OxiTop
IS 12-6 system (Wisenschaftlich-Technische Werkstätten (WTW), Weilheim,
Germany), a manometric measurement device. The measurement principle is
based on measuring pressure differences estimated by piezoresistive electronic
pressure sensors. Nitrification was suppressed by adding 0.05 ml of 5 g/l
N-Allylthiourea (WTW Chemical Solution No. NTH 600) solution per 50 ml
of sample water.

Concerning the analysis of nutrients in the liquid phase, oxidised aqueous
nitrogen was determined in all water samples as the sum of nitrate-nitrogen and
nitrite-nitrogen. However, nitrite-nitrogen concentrations were significantly low
(data not shown). Nitrate was reduced to nitrite by cadmium and determined as an

azo dye at 540 nm (using a Perstorp Analytical EnviroFlow 3000 flow injection analyser) following diazotisation with sulfanilamide and subsequent coupling with N-1-naphthylethylendiamine dihydrocloride (Allen, 1974).

Ammonia-nitrate and ortho-phosphate-phosphorus were determined by automated colorimetry in all water samples from reaction with hypochlorite and salicylate ions in solution in the presence of sodium nitrosopentacyanoferrate, and reaction with acidic molybdate to form a phosphomolybdenum blue complex, respectively (Allen, 1974). The coloured complexes formed were measured spectrometrically at 655 and 882 nm, respectively, using a Bran and Luebbe autoanalyser (Model AAIII).

A Whatman PHA 230 bench-top pH meter (for control only), a Hanna HI 9142 portable waterproof dissolved oxygen (DO) meter, a HACH 2100N turbidity meter and a Mettler Toledo MPC 227 conductivity, total dissolved solids (TDS) and pH meter were used to determine pH, DO, conductivity, turbidity and TDS, respectively. An ORP HI 98201 redox potential meter with a platinum tip electrode HI 73201 was used to measure pH. Composite water samples were analysed. All other analytical procedures were performed according to the American standard methods (Clesceri et al., 1998).

Concerning the analysis of major nutrients in Ecosoil (aggregate supplied by Atlantis Water Management Ltd.), 2 ml sulphuric acid (strength of 98%, v/v) and 1.5 ml hydrogen peroxide (strength of 30%, v/v) were used as an extraction media (Allen, 1974). Approximately 0.1 g of each dried sample and the associated digestion media were placed in a tube, and heated at 320°C for 6 h. Aliquots were taken and digests were made up to 100 ml with distilled water.

For analysis of total nitrogen, the following procedure was adopted: Ammonium (present in the digest) reacts with hypochlorite ions generated by alkaline hydrolysis of sodium dichloroisocyanurate. The reaction forms monochloroamine, which reacts with salicylate ions in the presence of sodium nitroprusside to form a blue indephenol complex. This complex is measured colorimetrically at 660 nm using a Bran and Luebbe autoanalyser (Model AAIII).

For analysis of total phosphorus, the following procedure was used: ortho-phosphate (present in the digest) reacts with ammonium molybdate in the presence of sulphuric acid to form a phosphomolybdenum complex. Potassium antimonyl tartrate and ascorbic acid are used to reduce the complex, forming a blue colour, which is proportional to the total phosphorus concentration. Absorption was measured at 660 nm using a Bran and Luebbe autoanalyser (Model AAIII).

For the analysis of total potassium, the digest was analysed by a flame atomic absorption spectrometer (Unicam 919, Cambridge, UK) at a wavelength of 766.5 nm and with a bandpass of 1.5 nm. Standards were prepared in 100 ml flasks using 2 ml concentrated sulphuric acid and 1.5 ml hydrogen peroxide (30% v/v), and made up to mark with de-ionised water. Caesium at a concentration of 100 mg/l was added to both standards and digests to overcome ionisation.

Metal concentrations were determined in the raw gully pot liquor and the outflow waters from the experimental rig on 16 June 2004. Samples for metal determinations were stored at $-19°C$ until analysis.

Concerning the analysis of Ecosoil and grass cuttings, composite samples were collected and stored at $-10°C$ prior to analysis. After thawing, approximately 2.5 g of each sample was weighed into a 100-ml digestion flask to which 21 ml of hydrochloric acid (strength of 37%, v/v) and 7 ml of nitric acid (strength of 69%, v/v) were added. The mixtures were then heated on a Kjeldahl digestion apparatus (Fisons, UK) for at least 2 h. After cooling, all solutions were filtered through a Whatman Number 541 hardened ashless filter paper into 100-ml volumetric flasks. After rinsing the filter papers, solutions were made up to the corresponding mark with deionised water. The method was adapted from the section on 'Nitric Acid–Hydrochloric Acid Digestion' (Clesceri et al., 1998).

An Inductively Coupled Plasma Optical Emission Spectrophotometer (ICP-OES) called TJA IRIS was supplied by ThermoElemental (USA), and used to analyse selected wastewater, Ecosoil and grass cutting samples. The purpose was to economically screen samples to determine various trace element concentrations and potential contaminants. Analytical precision (relative standard deviation) was typically between 5 and 10% for three individual aliquots.

27.4 Results and discussion

27.4.1 Comparison of costs

The overall capital and maintenance costs were calculated for each detention system for the first year of operation. Maintenance included litter removal and grass cutting, and was based on an area of 1000 m^2. Material prices were requested for a volume of 100 m^3 per aggregate to obtain realistic figures for a scaled-up detention system (industrial operation size). The five system configurations have standardised cost ratios of approximately 1.0:1.1:1.2:1.3:1.6 based on Edinburgh prices in November 2005. However, the actual prices are subject to negotiation (e.g., quantities ordered) and fluctuation on the market.

27.4.2 Inflow water quality

Table 27.3 summarises the inflow water quality. The standard deviations (data not shown) for all inflow parameters (except for DO, pH and temperature) are high due to the random selection of gully pots and seasonal variations (Butler and Parkinson, 1997; Scholz, 2004a).

In general, the gully pot liquor was more polluted in winter in comparison to the other seasons. There are various reasons for this including that the lower

Table 27.3 Gully pot liquor (inflow to systems): water quality variables (20/03/04–19/03/05)

Variable	Unit	Total no.	Mean spring[a]	Mean summer[b]	Mean autumn[c]	Mean winter[d]
BOD[e]	mg/l	78	33.5	29.0	31.8	27.6
Nitrate-N[f]	mg/l	71	0.5	1.8	0.7	0.3
Ammonia-N	mg/l	74	2.3	1.6	2.0	1.5
O-phosphate-P	mg/l	73	0.1	0.2	0.3	0.7
Suspended solids	mg/l	77	345.4	117.3	150.9	258.0
Total solids	mg/l	78	619.7	451.5	381.8	1604.7
Turbidity	NTU	84	89.2	57.4	67.2	126.2
Dissolved oxygen	mg/l	82	3.1	3.3	3.8	5.1
pH	–	84	6.79	7.16	7.02	7.38
Redox potential	mV	84	102.2	242.9	228.0	181.0
Conductivity	μS	83	285.8	142.0	149.0	239.3
Temperature (air)	°C	84	16.2	19.3	11.1	8.1
Temperature (gully pot)	°C	84	14.6	19.1	9.6	4.9

[a] 20/03-20/06/04; [b] 21/06-21/09/04; [c] 22/09/04–20/12/04; [d] 21/12/04–19/03/05; [e] 5-day @ 20°C N-Allylthiourea biochemical oxygen demand; [f] includes nitrite-nitrogen.

temperature in winter compared to other seasons results in a slower biodegradation rate within the gully pot (Table 27.3). Moreover, the retention time of the gully pot liquor in winter is likely to be shorter than in the remaining seasons due to more frequent rainfall events. A shorter retention time correlated positively with a lower biodegradation rate (Butler and Davies, 2000; Clesceri et al., 1998; Scholz, 2004a).

27.4.3 Comparison of outflow water quality

The overall filtration performance figures are summarised in Table 27.4, which should be compared with Table 27.3. Reduction efficiencies for BOD and SS (Table 27.4) are comparable to findings reported elsewhere (Bulc and Slak, 2003; Scholz, 2004a) for highway runoff treatment with constructed wetlands. The reductions of BOD (Table 27.4) were acceptable for most systems, if compared to minimum American and European standards for the secondary treatment of effluent.

Biochemical oxygen demand in contrast to SS (Table 27.4) outflow concentrations did not exceed the USA threshold of 30 mg/l (Tchobanoglous et al., 2003). However, some European standards or those of individual regional agencies (Cooper et al., 1996; Lim et al., 2003; Shutes et al., 2001) are more stringent; e.g., BOD <20 mg/l. The BOD outflow concentration was also lower than the UK standard (Scholz, 2004a) for secondary treated wastewater of 20 mg/l (Table 27.4).

The mechanisms of improving the water quality are similar in all systems. The wetland filters on top of the detention cells retain solids, and subsequently

Table 27.4 Relative change (%) of outflow variables (20/03/04–19/03/05)

Var.	System 1 SP[a]	System 1 SU[b]	System 2 SP[a]	System 2 SU[b]	System 3 SP[a]	System 3 SU[b]	System 4 SP[a]	System 4 SU[b]	System 5 SP[a]	System 5 SU[b]
BOD[c]	93	90	94	89	88	90	91	91	94	91
NO_3[d]	−1479	−1001	−829	−1345	−481	−496	−562	−825	−850	−190
NH_4[e]	74	84	86	87	78	86	76	85	82	82
PO_4[f]	16	−133	12	−131	12	−101	8	−168	2	−208
SS[g]	28	74	31	80	32	79	31	79	25	79
Turb[h]	91	91	93	91	90	89	80	91	85	60

Var.	System 1 AU[i]	System 1 WI[j]	System 2 AU[i]	System 2 WI[j]	System 3 AU[i]	System 3 WI[j]	System 4 AU[i]	System 4 WI[j]	System 5 AU[i]	System 5 WI[j]
BOD[c]	94	92	89	94	94	87	92	93	93	88
NO_3[d]	−2374	−834	−2191	−1050	−731	−66	−1238	−472	−472	−363
NH_4[e]	63	77	93	84	65	71	97	38	93	32
PO_4[f]	−3	75	22	77	30	80	18	80	33	76
SS[g]	97	97	96	98	97	98	97	97	96	98
Turb[h]	96	96	96	95	93	93	93	95	94	94

[a]mean of spring (20/03/04–20/06/04); [b]mean of summer (21/06/04–21/09/04) [c]5-day @ 20°C N-Allylthiourea biochemical oxygen demand (mg/l); [d]nitrate-nitrogen (mg/l); [e]ammonia-nitrogen (mg/l); [f]otho-phosphate-phosphorus (mg/l); [g]suspended solids (mg/l); [h]turbidity (NTU); [i]mean of autumn (22/09/04–20/12/04); [j]mean of winter (21/12/04–19/03/05).

degrade organic matter. However, the greatest treatment performance is achieved due to extended storage within the detention cells that function similar to a covered wastewater treatment pond.

A regression analysis has shown that BOD, ammonia-nitrogen, nitrate-nitrogen and ortho-phosphate–phosphorus can be estimated with conductivity and total dissolved solids using a second-order polynomial equation. For example, BOD, nitrate-nitrogen and ammonia-nitrogen can be determined with conductivity; the corresponding coefficients of determination (r^2) for Filter 4, for example, are 0.60, 0.71 and 0.76, respectively. This would result in the reduction of costs and sampling effort. However, statistical relationships between other variables were not significant.

Furthermore, it has been suggested that mature and viable microbial biomass, in contrast to aggregates with high adsorption capacities (e.g., Ecosoil) and turf, is responsible for the high overall filtration performances (Cooper et al., 1996; Scholz and Martin, 1998a). However, it is difficult to classify objectively a biological system as mature without having undertaken detailed microbiological work.

Finally, analysis by ICP-OES of selected inflow and outflow samples for a suite of cations showed that all waters generally contained low concentrations of heavy metals. Measured elemental concentrations were either low (barium, calcium, magnesium and manganese) or close to the detection limit (iron), and for most heavy metals (including aluminum, copper and cadmium) below the detection limit. Dissolved zinc was the pollutant measured with the highest mean concentration. The mean inflow concentration for zinc was 0.14 mg/l, and the corresponding outflow concentrations were 0.07 mg/l (SD of 0.05 mg/l).

27.4.4 Ecosoil and turf

Ecosoil did not contribute to elevated nutrient concentrations due to very low total nitrogen, total phosphorus and total potassium concentrations of 65, 46 and 1367 mg/kg, respectively. A recent soil quality analysis for areas in Glasgow, where SUDS were considered for implementation, indicated total nitrogen, total phosphorus and total potassium concentrations of 1612, 605 and 4562 mg/kg, respectively (Scholz et al., 2005). It follows that Ecosoil does function only as a very weak fertiliser, and that it is therefore unlikely to contribute to eutrophication after the release of the treated stormwater to the nearby watercourse.

Furthermore, Ecosoil contained only trace amounts of heavy metals (except for aluminum): 1036, 24 and 7 mg/kg dry weight of aluminum, zinc and nickel, respectively. All other metal concentrations were below the detection limit of the instrument. However, even the aluminum concentrations are similar to values reported elsewhere for urban soil (Scholz et al., 2005).

The influence of turf (System 5) on the organic matter content of the outflow was studied. The BOD and SS concentrations within the outflow from the planted system compared to the unplanted gravel and sand systems were similar (Tables 27.3 and 27.4).

Moreover, grass on top of Filter 6 was cut when the length was ≥10 cm for optical reasons and to reduce the overall nutrient load. Total nitrogen, total phosphorus and total potassium concentrations were 3001, 640 and 6909 mg/kg fresh weight. The presence and associated harvesting of grass seemed to have a positive effect on the overall nitrate-nitrogen outflow concentration that was lower for System 5, if compared to the remaining systems (Tables 27.3 and 27.4).

27.5 Conclusions and further research

The BOD outflow concentrations were mostly below the USA threshold of 30 mg/l for secondary treated wastewater. The below-ground stormwater detention system did show signs of overloading resulting in occasionally relatively high SS and nitrate-nitrogen concentrations. It follows that further treatment would be required. Moreover, denitrification was not completed, and longer retention times are therefore suggested. Nitrate-nitrogen was lower in the outflow of the planted system (turf on the top).

Hydraulic gully pot liquor (concentrated stormwater runoff) loads exceeded three times the mean annual rainfall for all systems. Therefore, it is likely that the SS concentration would be much lower in the field under 'real' conditions.

An analysis of variance indicated that there was no significant difference between most systems in terms of their treatment performance (e.g., BOD and SS) despite of their different set-ups. It follows that all systems function as covered wastewater stabilization ponds regardless of the corresponding soil filter type.

Sampling costs and effort can be reduced using relationships derived from regression analysis between expensive variables that can be substituted by low-cost ones. For example, BOD can be replaced by conductivity for internal control purposes.

Ecosoil did contain relatively low concentrations of nutrients and metals (except for aluminum). It follows that higher investment costs for more complex systems are not justified based on a water quality analysis alone.

However, further research related to the potential hydraulic and structural benefits of additional aggregates such as Ecosoil are required. It should be assessed if Ecosoil can help to evenly distribute the inflow water, and if it is capable of supporting 'moving' and variable heavy loads on paved areas.

Chapter 28

Modeling of constructed wetland performance

28.1 Summary

K-nearest neighbours (KNN), support vector machine (SVM) and self-organizing map (SOM) were applied to predict five-day @ 20°C N-Allylthiourea biochemical oxygen demand (BOD) and suspended solids (SS), and to assess novel alternative methods of analyzing water quality performance indicators for constructed treatment wetlands (Chapter 26). Concerning the accuracy of prediction, SOM showed a better performance compared to both KNN and SVM.

Moreover, SOM had the potential to visualize the relationship between complex biochemical variables. However, optimizing the SOM requires more time in comparison to KNN and SVM, because of its trial and error process in searching for the optimal map.

The results suggest that BOD and SS can be efficiently estimated by applying machine learning tools with input variables such as redox potential and conductivity, which can be monitored in real time. Their performances are encouraging and support the potential for future use of these models as management tools for the day-to-day process control.

28.2 Introduction

28.2.1 Project purpose

The purpose of this study was to examine the goodness of applying KNN, SVM and SOM to predict the outflow water quality of experimental constructed treatment wetlands by comparing the accuracy of these models. Additionally, this study describes how machine learning can be used for water treatment monitoring and assessment.

The objectives are to assess the following:

- The most appropriate method for assessing the performance of constructed treatment wetlands, considering both the accuracy of estimations and input costs;
- The potential of KNN, SVM and SOM for analyzing biochemical performance data;
- The optimization of input variables associated with predictive models; and
- The potential use of KNN, SVM and SOM as teaching tools to enhance the understanding of 'black box' systems.

28.2.2 Machine learning applied to wastewater treatment processes

Constructed treatment wetlands are often seen as complex 'black box' systems, and the processes within an experimental wetland are difficult to model due to the complexity of the relationships between most water quality variables (Gernaey et al., 2004). However, it is necessary to monitor, control and predict the treatment processes to meet environmental and sustainability policies, and regulatory requirements such as secondary wastewater treatment standards (Scholz, 2004a,b).

The measurement of BOD and SS concentrations is widely applied for wastewater before and after treatment, as they give a general indication of the water quality status. However, taking BOD measurements can both be expensive (measurements are labor intensive and capital costs of modern online equipment are relatively high; approximately £15 000) and only of historical value (results are not available until five days after the sample has been taken). Therefore, an indirect method of prediction of BOD and SS, if it could be made reliable enough, would be advantageous.

A variety of machine learning methods such as KNN and artificial neural networks (ANN) have been widely used in a broad range of domains including wastewater treatment engineering. The KNN technique is based on a simple methodology and a memory-based model defined by a set of examples for which the outcomes are known. Moreover, the KNN model estimates the outcome by finding k examples that are closest in distance to the target point. Thus, the determination of the optimal value for k is essential in building the KNN model, because it should be the maximum number of neighbors with the minimum possible error (Ruiz-Jimenez et al., 2004).

The KNN model has been compared with advanced neural networks, and tested for a wide range of areas such as medical diagnosis, chemical analysis and remote sensing (Carpenter and Markuzon, 1998; Dong et al., 2005; Ruiz-Jimenez et al., 2004). In the case of the application of KNN models in the wastewater treatment industry, Belanche et al. (2000) employed a KNN model for reference purposes in predicting sludge bulking.

Neural networks are relatively effective in simulating and predicting water treatment processes. The advantages of ANN are as follows: ease of use, rapid prototyping, high performance, minor assumptions, reduced expert knowledge required, non-linearity, multi-dimensional and easy interpretation (Werner and Obach, 2001).

Artificial neural networks such as feed-forward neural networks were developed to predict the effluent concentrations including BOD, chemical oxygen demand (COD), and SS for wastewater treatment plants (Grieu et al., 2005; Hamed et al., 2004; Onkal-Engin et al., 2005), and to control water treatment processes automatically; e.g., by modeling the alum dose (Maier et al., 2004). These studies have shown that ANN could be applied to establish a mathematical relationship between variables describing a process state and different measured quantities.

Although ANN methods are cost-effective and highly reliable in analyzing processes, the traditional neural networks have suffered from their inherent drawbacks; i.e. over-training, local minima, poor generalization and difficulties in their practical application (Lu and Wang, 2005). The SVM, a supervised machine learning technique, developed by Vapnik (1995), provides a novel approach to improve the generalization performance of neural networks.

Originally, SVM models have been applied for pattern recognition problems. However, along with the introduction of Vapnik's ε intensive loss function, SVM have also been extended to solve non-linear regression estimation problems (Vapnik, 1995; Pai and Hong, 2005). It classifies the data based on the similarity between the examples measured by the similarity function or kernel function. This function can be chosen according to the problem at hand, and thus making the algorithm flexible in handling a wide variety of problems (Dubey et al., 2005). Moreover, previous studies demonstrated that the SVM is superior to the conventional neural network in predicting chemical and biological variables (Liu et al., 2004; Lu and Wang, 2005). However, this novel method has not yet been applied in the field of wastewater treatment including constructed treatment wetlands.

The SOM, which is based on an unsupervised learning algorithm, uses powerful pattern recognition analysis and clustering methods, and at the same time provides excellent visualization capabilities (Garcia and Gonzalez, 2004). The SOM is able to map a structured, highly dimensional data set onto a much lower dimensional network in an 'orderly' fashion (Lu and Lo, 2004). It offers the distinctive ability to gather knowledge by detecting the patterns and relationships from a given data set, learning from relationships and adapting to change. The SOM potentially outperforms current methods of analysis, because it can successfully deal with the non-linearity of a system, handle 'noisy' or irregular data, and be easily updated (Hong et al., 2002).

Interesting approaches of SOM have been reported in water quality assessment. The SOM models were developed to evaluate the state of water quality of a reservoir, and to predict the trophic status of coastal waters, showing a strong ability to identify the diversity between data (Aguilera et al., 2001; Gervrey et al., 2004).

Moreover, Verdenius and Broeze (1999) used an SOM model as an indexing mechanism in case-based reasoning algorithms to control wastewater treatment processes, and it was employed to diagnose the diverse states of a wastewater treatment plant (Garcia and Gonzalez, 2004; Hong et al., 2003). These studies demonstrated that the SOM can assist a process engineer by analyzing multi-dimensional data and simplifying them into visual information, which can be easily applied to control plant performance. However, applications of SOM in water treatment process control are relatively new and were not implemented as much as traditional neural networks such as the free forward neural networks (Grieu et al., 2005; Hamed et al., 2004).

It follows that comparative studies of traditional KNN models with novel neural networks (e.g., SVM and SOM) applied to predict wastewater treatment performances are required to advance operation process control. Moreover, ANN should be used to find out if these models can be effectively applied to predict water quality variables such as BOD and SS effluent concentrations in wastewater treatment systems, using their potential for data classification and clustering.

28.3 Methodology and software

28.3.1 Experimental data and variables

Twelve wetland filters were operated to assess the system performance concerning the treatment of gully pot liquor in a cold climate. Gully pot liquor is concentrated surface runoff, which is detained in the wet gully pot until it overflows into the sewer due to incoming surface runoff from new rainfall events. Design, operation and monitoring methods of the system were explained previously (Scholz, 2004a,b). Different packing order arrangements of filter media and plant roots were used in the wetland filters as described previously by Scholz (2004a). In comparison, Filters 3, 5, 7 and 9 are similar to gravel and slow sand filters, and Filters 4, 6, 8 and 10 are typical reed bed filters. Filters 11 and 12 are more complex in their design and operation.

Experimental data were collected by monitoring the effluent concentrations of the filters including BOD and SS for more than two years (09/09/02–21/09/04). The number of data were comparable to those used in other prediction models (Aguilera et al., 2001; Liu et al., 2004). These data were stored in the data base together with up to six input variables; turbidity (NTU), conductivity (μS), redox potential (mV), outflow water temperature ($°C$), dissolved oxygen, DO (mg/l) and pH (-). The corresponding output variable was either BOD (mg/l) or SS (mg/l). The input variables were selected according to their goodness of correlation with both BOD and SS (Scholz, 2003), because they are relatively cost-effective and easy to measure.

28.3.2 k-nearest neighbours

A KNN model used to predict the effluent BOD and SS concentrations of the wetland systems were created using MATLAB 7.0. Each KNN model is based on the mean of the outcomes of the k-nearest neighbours. The local similarity (i.e. the similarity of a past case and the problem case with respect to only one variable) is found via a mathematical function of the difference between each past case and a problem case.

The Gaussian function (bell-shaped curve) used to map the local difference onto the local similarity is defined in Eq. (28.3.1), which applies fuzzy theory (Dubois and Prade, 1998). This function has a tuning parameter, which is used to determine the flatness of the smoothing function.

$$f(x) = e^{\left(-0.5\left(\frac{x}{\alpha \times SDV_i}\right)^2\right)}$$

(28.3.1)

where:

f   = the function, which converts the local difference into the local similarity;

x   = the local difference between each past case and a problem case;

α = the tuning parameter; and

SDV_i = the standard deviation of the local differences of variable i.

The global similarity, which is the similarity between the past case and the problem case, considering all variables of a past case, can be found from the local similarity of each variable. Each local similarity is first multiplied by a weighting factor that corresponds to the importance of that variable in predicting the output.

An algorithm proposed by Duch and Grundzinski (1999) was used to identify feature weightings. For the initial ranking of features, all weighting factors are set to one, and evaluation with a single feature turned off (i.e. set to zero) is made for all features. Thus, ranking is done in the same way as the feature dropping selection method (Duch and Grudzinski, 1999). The important feature has a fixed weighting factor of one, and the optimal weighting value for the second factor in the ranking is determined by the search procedure. The remaining factors are all fixed to one. The search is implemented by means of the leave-one-out (LOO) cross validation process (Kohavi, 1995).

When the global similarity of each past case with the problem case is found, the past cases can be selected by the first k closest cases. The tuning parameter α of the Gaussian function and the k value were determined by LOO cross validation in the training phase (Duch and Grudzinski, 1999).

28.3.3 Support vector machine

Concerning SVM, the basic idea is to map original data into a feature space, which has a large number of dimensions via a non-linear mapping function $\phi_i(x)$, that is usually unknown, and then carry out linear regression in the feature space (Vapnik, 1995). Hence, the regression addresses a problem of estimating a function based on a given data set (Eq. (28.3.2)).

$$G = \{(x_i, y_i)\}_i^1 \qquad (28.3.2)$$

where:

G = data set;
x_i = input vector; and
y_i = desired values, which are produced from the non-linear mapping function $\phi_i(x)$.

The SVM approximates the optimum decision function using Eq. (28.3.3):

$$f(x) = \sum_{i=1}^{1} w_i \phi_i(x) + b \qquad (28.3.3)$$

where:

$f(x)$ = decision function;
$\phi_i(x)$ = non-linear mapping function representing the features of inputs; and
w and b = coefficients, which are estimated by minimizing the regularized risk function (see below).

The regularized risk function $R(C)$ is shown in Eq. (28.3.4).

$$R(C) = C \frac{1}{1} \sum_{i=1}^{1} L_\varepsilon(y_i, f(x_i)) + \frac{1}{2} \|w\|^2 \qquad (28.3.4)$$

where:

$R(C)$ = regularized risk function;
C = regularized constant determining the trade-off between the training error and the model flatness;

$\frac{1}{1} \sum_{i=1}^{1} L_\varepsilon(y_i, f(x_i))$ = empirical error measured by the ε-insensitive loss function;

y_i = desired values;
$f(x_i)$ = decision function; and
$\frac{1}{2} \|w\|^2$ = measurement of function flatness.

The ε-insensitive loss function is defined in Eq. (28.3.5).

$$L_\varepsilon\left(y_i, f(x_i)\right) = \begin{cases} |y_i - f(x_i)| - \varepsilon, & |y_i - f(x_i)| \geq \varepsilon \\ 0 \end{cases} \tag{28.3.5}$$

where:

L_ε = ε-insensitive loss function;
y_i = desired values;
$f(x_i)$ = decision function; and
ε = prescribed parameter.

By introducing the kernel function, Eq. (28.3.3) can be transformed into the explicit Eq. (28.3.6).

$$f(x) = \sum_{i=1}^{1} \left(\alpha_i - \alpha_i^*\right) K\left(x_i, x_j\right) + b \tag{28.3.6}$$

where:

$f(x)$ = decision function;
α, α_i = Lagrange multipliers; and
$K(x_i, x_j) = \phi(x_i) \times \phi(x_j)$ = kernel function in the feature space.

For the kernel function, there are several design choices such as the linear, polynomial and radial basis functions (RBF), and the sigmoid kernel. However, most of the previous research selected the RBF kernel, which non-linear maps samples into a higher dimensional space, unlike the linear kernel (Dong et al., 2005). The RBF kernel is shown in Eq. (28.3.7).

$$K\left(x_i, x_j\right) = \exp\left\{-\gamma \left\|x_i - x_j\right\|^2\right\} \tag{28.3.7}$$

where:

$K\left(x_i, y_i\right)$ = kernel function; and
γ = kernel parameter.

More detailed theory on the SVM can be found in Vapnik (1995). In this study, the SVM[light] was used due to its fast optimization algorithms and good potential for regression (Joachims, 1999). Concerning the kernel function, the RBF kernel was selected to analyze the cases, which show a non-linear relationship between input and output data set in this study. The RBF kernel contains the parameters γ, C and ε (see above). There are no general rules determining these parameters

(Lu and Wang, 2005). A five-fold cross validation was conducted to find out appropriate parameters for training steps (Dong et al., 2005).

28.3.4 Self-organizing map

A SOM consists of neurons, which are connected to adjacent neurons by neighbourhood relations. In the training step, one vector x from the input set is chosen, and all the weight vectors of the SOM are calculated using some distance measure such as the Euclidian distance (Kohonen, 2001). The neuron, whose weight vector is closest to the input x is called the best-matching unit (BMU), subscripted here by c (Eq. 28.3.8):

$$\|x - m_c\| = \min\{\|x - m_i\|\} \qquad (28.3.8)$$

where:

x   = input vector;
m   = weight vector; and
$\|\ \|$ = distance measure.

After finding the BMU, the weighting vectors of the SOM are updated so that the BMU is moved closer to the input vector. The SOM update rule for the weight vector of the unit is shown in Eq. (28.3.9). The detailed algorithm of the SOM can be found in Kohonen (2001) for theoretical considerations:

$$m_i(t+1) = m_i(t) + \alpha(t)\,h_{ci}(t)\,[x(t) - m_t(t)] \qquad (28.3.9)$$

where:

m(t)   = weight vector indicating the output unit's location in the data space at time t;
$\alpha(t)$ = the learning rate at time t;
$h_{ci}(t)$ = the neighbourhood kernel around the 'winner unit' c; and
x(t)   = an input vector drawn from the input data set at time t.

After the SOM has been trained, the map needs to be evaluated to find out if it has been optimally trained, or if further training is required. The SOM quality is usually measured with two criteria: quantization error (QE) and topographic error (TE). The QE is the average distance between each data point and its BMU, and TE represents the proportion of all data for which the first and second BMU are not adjacent with respect to the measurement of topology preservation (Kohonen, 2001).

In this study, the SOM toolbox (Version 2) for Matlab 5.0 developed by the Laboratory of Information and Computer Science at the Helsinki University of

Technology was used (Vesanto et al., 1999). After training the map with different map sizes, the optimum map size was determined on the basis of the minimum QE and minimum TE. The prediction was implemented by finding BMU in the map for each test data set.

28.4 Results and discussion

28.4.1 Performance evaluation

When comparing the performances of different models, the scale-dependent measures based on the absolute error or squared error have been commonly used (Gevrey et al., 2004; Maier et al., 2004). In this study, the performances of each model were measured by the mean absolute scaled error (MASE) method proposed by Hyndman and Koehler (2005), because it is independent of the scale of the data, less sensitive to outliers, more easily interpreted and less variable for small samples compared to most of the other methods (see above). The MASE is defined in Eq. (28.4.1).

$$\text{MASE} = \frac{\text{MAE}}{\dfrac{1}{n-1}\sum_{i=2}^{n}|m_i - m_{i-1}|} \tag{28.4.1}$$

where:

MASE = mean absolute scaled error;

$\text{MAE} = \dfrac{1}{n}\sum_{i=1}^{n}|m_i - p_i|$;

M_i = measured values;

p_i = predicted values; and

n = number of data sets.

28.4.2 Correlation analysis

Table 28.1 summarizes the findings from a correlation analysis comprising input (turbidity, conductivity, redox potential, outflow water temperature, DO and pH) and target (BOD or SS) variables. Correlations were all weak, except between BOD and turbidity, SS and turbidity, and SS and conductivity (all at 1% significance level). Therefore, turbidity and conductivity are likely to be the most important input variables.

Table 28.1 Correlation coefficients from a correlation analysis comprising input (column headings) and target (row headings) variables

Variable	Turbidity (NTU)	Redox potential (mV)	pH (−)	Conductivity (μS)	Temperature (°C)	Dissolved oxygen (mg/l)
BOD[a] (mg/l)	0.413	−0.338	−0.271	0.254	−0.120	−0.074

Variable	Conductivity (μS)	Turbidity (NTU)	Temperature (°C)	Redox potential (mV)	Dissolved oxygen (mg/l)	pH (−)
SS[b] (mg/l)	0.930	0.509	−0.322	−0.308	−0.127	0.013

[a]five-day @ 20°C N-Allylthiourea biochemical oxygen demand; [b]suspended solids.

28.4.3 Optimization of input variables

When analyzing the data set with the KNN model, the optimal k value and weighting factors for all variables were determined by LOO cross validation. The selected k value was between 3 and 5 for most data sets, and weighting factors were different depending on the individual characteristics of the chosen case base.

For example, when predicting the outflow BOD concentration for Filter 8, the k value was fixed at 3 and the weighting factors 1.00, 0.78, 1.00, 0.30 and 0.78 for water temperature (°C), redox potential (mV), conductivity (μS), pH (−) and turbidity (NTU), respectively, of the outflow were assigned.

When conducting the SVM analysis, the parameters C and ε were identified by the five-fold cross-validation approach (see above). Theoretically, a small value of C will under-fit the training data, because the weight placed on the training data is too small, thus resulting in a large error for the test data set. On the contrary, when C is too large, the SVM model will be over-trained (Dong et al., 2005). The ε is set to be 0.1 when varying C in the one-time search method (Cao et al., 2003). There exists an optimum point for C, as shown in Fig. 28.1a. Thus C was determined to be 10 at the point of the lowest mean absolute error (MAE) value.

In general, the larger ε, the smaller is the number of support vectors, and thus the sparser the representation of the solution. However, if ε is too large, it can deteriorate the accuracy of the training data set (Cao et al., 2003). According to Fig. 28.1b, parameter ε was fixed at 0.1. Furthermore, γ of the RBF kernel was set to 0.2 according to Eq. (28.4.2) as discussed elsewhere (Chang and Lin, 2005; Dong et al., 2005).

$$\gamma = \frac{1}{n} \qquad (28.4.2)$$

where:

γ = parameter of the kernel function; and
n = number of variables in the input data set.

In the SOM model, the map size is the most important factor in detecting the differences of data. If the map is too small, it might not explain some important differences. On the contrary, it is possible to over-train the models (Leflaive et al., 2005). After creating maps with several different map sizes, the optimum map size, which has lower errors for both QE and TE, was chosen. For example, when

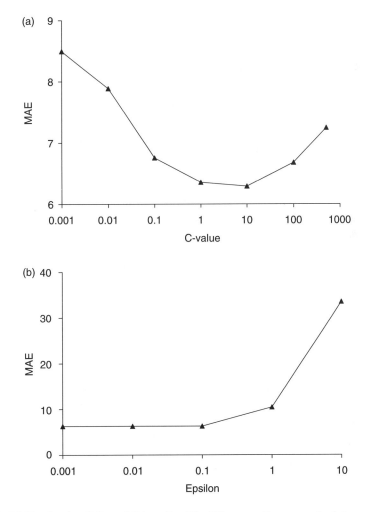

Fig. 28.1 (a) C-value ($\varepsilon = 0.1$), and (b) epsilon (C = 10) versus the mean absolute error (MAE) of the training set based on five-fold cross-validation.

Table 28.2 Quantization error (QE) and topographic error (TE) for different self-organizing map sizes

Map size	608	1508	2006	2304	2613	2911
QE	0.038	0.017	0.008	0.007	0.005	0.004
TE	0.108	0.091	0.124	0.091	0.148	0.120

predicting the BOD of Filter 8, the map size comprised 2304 units as shown in Table 28.2.

Table 28.3 summarizes the findings of an input variable combination optimization exercise. The purpose was to identify the optimum combination of input variables by estimating both BOD and SS with as few input variables as possible to reduce the prediction error, and sampling costs and effort. Therefore, the data set contained the following input variables in order of their priority based on a correlation analysis (Table 28.1): turbidity, redox potential, pH, conductivity and outflow water temperature in terms of their correlation with BOD; conductivity, turbidity, outflow water temperature, redox potential and DO in terms of their correlation with SS.

Table 28.3 Optimization of input attribute combinations for Filter 8: k-nearest-neighbours (KNN), support vector machine (SVM) and self-organizing map (SOM) applied to predict the five-day @ 20°C N-Allylthiourea biochemical oxygen demand (BOD) (mg/l) and the suspended solids (SS) (mg/l) concentrations

Input variables	Number of test data sets	KNN		SVM		SOM	
		MASE[a]	r^{2b}	MASE[a]	r^{2b}	MASE[a]	r^{2b}
		BOD prediction					
1	60	1.42	0.394	1.35	0.621	0.29	0.998
1+2	60	1.49	0.394	1.27	0.621	0.30	0.988
1+2+3	60	1.51	0.330	1.09	0.639	0.34	0.984
1+2+3+4	60	1.34	0.342	0.95	0.699	0.46	0.955
1+2+3+4+5	60	1.01	0.680	0.94	0.720	0.41	0.947
		SS prediction					
4	64	0.55	0.936	0.38	0.959	0.29	0.999
4+1	64	0.44	0.943	0.40	0.950	0.30	0.992
4+1+5	64	0.37	0.948	0.39	0.950	0.31	0.980
4+1+5+2	64	0.39	0.955	0.39	0.954	0.31	0.969
4+1+5+2+6	64	0.39	0.951	0.36	0.949	0.39	0.956

[a] mean absolute scaled error; [b] prediction coefficient of determination.
Note: The training and test data sets contained the following input variables: 1 = turbidity (NTU); 2 = redox potential (mV); 3 = pH (-); 4 = conductivity (μS); 5 = outflow water temperature (°C); 6 = dissolved oxygen (mg/l). The figures for the best combinations of variables are underlined.

In the KNN model, best prediction results were obtained with five input variables for both BOD and SS. However, applying the SOM model, the BOD and SS predictions were most accurate with the single input variable turbidity and conductivity, respectively. The best combination of variables for each model is highlighted (underlined figures) in Table 28.3. Consequently, input variable combinations for each model were determined on the basis of this information.

28.4.4 Comparison of applications

The prediction results of the outflow BOD and outflow SS using KNN, SVM and SOM models are summarized in Table 28.4. Figures 28.2 and 28.3 visualize the

Table 28.4 Constructed treatment wetlands: k-nearest-neighbours (KNN), support vector machine (SVM) and self-organizing map (SOM) applied to predict the five-day @ 20°C N-Allylthiourea biochemical oxygen demand, BOD (mg/l), and the suspended solids, SS (mg/l), concentrations

Filter number	Number of test data sets	KNN		SVM		SOM	
		MASE[a]	r^{2b}	MASE[a]	r^{2b}	MASE[a]	r^{2b}
		BOD prediction					
3	55	1.02	0.461	0.98	0.384	0.77	0.997
4	58	1.01	0.681	1.07	0.713	0.17	0.994
5	59	0.58	0.696	0.67	0.666	0.35	0.994
6	61	0.76	0.798	0.83	0.643	0.44	0.997
7	58	0.78	0.609	0.77	0.503	0.85	0.999
8	60	1.01	0.680	0.94	0.720	0.29	0.998
9	57	0.76	0.544	0.90	0.431	0.20	0.998
10	60	0.71	0.744	0.87	0. 513	0.41	0.999
11	59	0.79	0.645	0.96	0.451	0.37	0.985
12	115	0.88	0.342	0.87	0.213	0.24	0.997
3–12	642	0.79	0.550	0.84	0.463	0.36	0.888
		SS prediction					
3	60	0.49	0.973	0.33	0.966	0.17	0.999
4	61	0.38	0.957	0.28	0.954	0.08	0.998
5	60	0.66	0.938	0.28	0.978	0.20	0.999
6	64	0.57	0.941	0.32	0.940	0.17	0.998
7	57	0.32	0.892	0.38	0.883	0.08	0.995
8	64	0.37	0.948	0.36	0.949	0.29	0.999
9	62	0.53	0.882	0.51	0.883	0.54	0.996
10	65	0.48	0.922	0.51	0.949	0.62	0.998
11	64	0.42	0.839	0.36	0.895	0.48	0.999
12	113	0.83	0.803	0.68	0.770	0.47	0.953
3–12	670	0.54	0.844	0.43	0.883	0.31	0.932

[a]mean absolute scaled error; [b]prediction coefficient of determination.

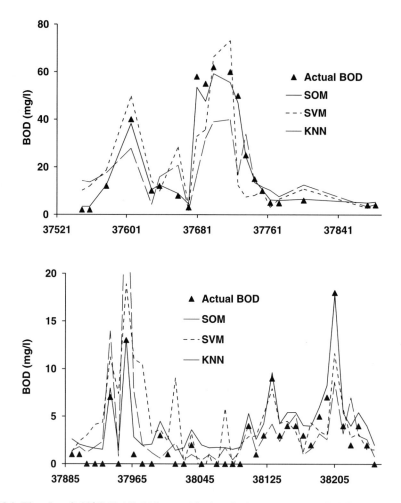

Fig. 28.2 Five-day @ 20°C N-Allylthiourea biochemical oxygen demand (BOD) predicted by k-nearest-neighbors (KNN), support vector machine (SVM) and self-organizing map (SOM) for (a) the first year (22/09/02 – 21/09/03), and (b) the second year (22/09/03 – 21/09/04) of wetland operation. The entry (32.6 mg/l on 03/12/03) for the KNN prediction is beyond the displayed range of BOD.

BOD and SS prediction performances of the KNN, SVM and SOM models for Filter 8 (typical UK reed bed according to Scholz (2004a,b)).

The results show that the BOD and SS concentrations are reasonably well predicted with given input variables. Despite the greater variability of SS in contrast to BOD (Scholz, 2004a,b), SS has a smaller MASE value between measured and predicted concentrations in comparison to BOD (Table 28.4).

The MASE from the BOD prediction with SOM, SVM and KNN are 0.36, 0.84 and 0.79, respectively. While the performance of the SVM model was not

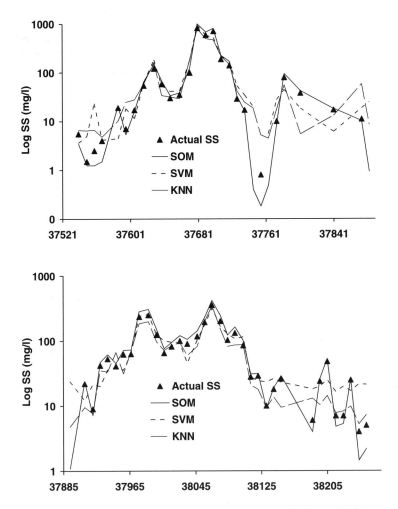

Fig. 28.3 Suspended solids (SS) curve predicted by k-nearest-neighbors (KNN), support vector machine (SVM) and self-organizing map (SOM) for (a) the first year (22/09/02 – 21/09/03), and (b) the second year (22/09/03 – 21/09/04) of wetland operation.

superior to that of KNN, the SOM model gave excellent prediction performance values compared to the other models (Table 28.5; Figs. 28.2 and 28.3).

To ensure the statistical significance of these findings, the prediction results were compared with each other, and assessed by using an analysis of variance (ANOVA). The ANOVA threshold for statistically significant findings is P < 0.05. It follows that pairs of MASE associated with P ≥ 0.05 can be regarded as similar. The analysis showed that the MASE of the SOM model were statistically different from those of SVM and KNN. However, the MASE of SVM and KNN were similar to each other, showing that the corresponding P-value was >0.05.

Table 28.5 Correct prediction of compliance (%) for the estimation of the five-day @ 20°C N-Allylthiourea biochemical oxygen demand, BOD (mg/l), and the suspended solids, SS (mg/l), concentrations

	SOM[a]	SVM[b]	KNN[b]
BOD	97.0	87.9	87.2
SS	96.4	90.0	88.1

[a] self-organizing map;
[b] support vector machine and;
[c] k-nearest neighbours.

From the findings, it can be concluded that the SOM model outperformed the KNN and SVM models.

Figure 28.4 shows the regression analysis between measured and predicted BOD, and measured and predicted SS for Filter 8 using the SOM model. The associated data set contained turbidity and conductivity for both BOD and SS predictions. The application of linear trendlines resulted in very good fits for both target variables; the prediction coefficients of determination are 0.998 for BOD and 0.999 for SS.

Concerning the supervised and unsupervised methods, a previous study has shown that the prediction accuracy of unsupervised neural networks is lower than the one for supervised networks, as generally expected (Lee et al., 2005a). However, this result indicated that the SOM model is superior to the SVM model. Moreover, the SOM model provides better prediction results with smaller input variables.

The outstanding performance of the SOM models is assumed to be attributed to the potential of clustering and classification of data. Particularly, in comparison with other neural networks, SOM was convenient for detecting the outliers, which are displayed in particular parts of the map without affecting the remaining parts, because each outlier takes its place in one unit of the map, and only the weights of that unit, and its neighbours, are affected (Cereghino et al., 2001; Leflaive et al., 2005).

Additionally, the SOM model showed its high performance in visualization with respect to the relationship for non-linear and complex biochemical data sets. Visualization gives better understanding of the relationships between most variables in biochemical processes. Figure 28.5 displays component planes on the trained map in gray scale.

The unified distance matrix (U-matrix) visualizes distances between neighbouring map units, and helps to identify the cluster structures of the map. Each component plane shows values for each variable with its corresponding unit. A cluster can be identified in the upper part of the U-matrix (Fig. 28.5). In the

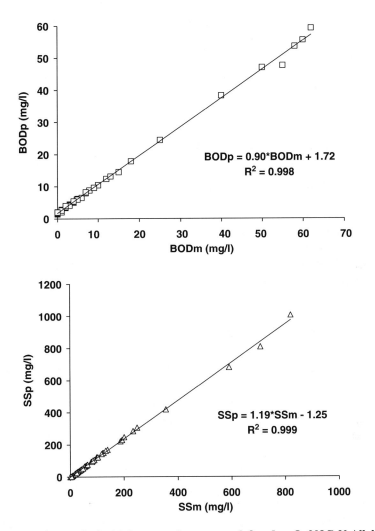

Fig. 28.4 Regression analysis (a) between the measured five-day @ 20°C N-Allylthiourea biochemical oxygen demand (BODm) and the predicted BOD (BODp) applying self-organizing map (SOM), and (b) between the measured suspended solids (SSm) and the predicted SS (SSp) applying the SOM model for Filter 8.

equivalent area of each component plane, the temperature and redox potential are high. On the other hand, the conductivity, turbidity and BOD concentrations are low in the upper part of the planes. Figure 28.5 indicates that high BOD is associated with high temperature, high redox potential, low conductivity and low turbidity.

The likelihoods of correct predictions, if the effluent concentrations are either below or above the thresholds for secondary wastewater treatment, are also

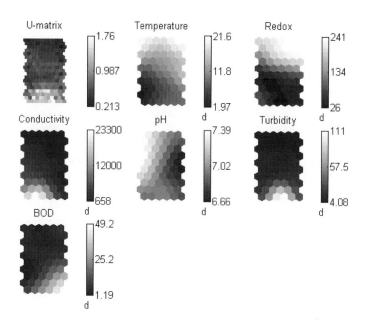

Fig. 28.5 Visualization of variables (outflow water temperature (°C), redox potential (redox, mV), conductivity (μS), pH (-), turbidity (NTU) and five-day @ 20°C N-Allylthiourea biochemical oxygen demand (BOD) (mg/l)) on the self-organizing map trained to predict the BOD of Filter 8. The map size is sixty units. The U-matrix is the unified distance matrix.

shown in Table 28.5. The BOD and SS concentrations for compliance are 20 and 30 mg/l, respectively (Scholz, 2003). The correct predictions of compliance were all >87%. The probabilities are therefore all at least by 0.37 higher in comparison to 'pure guessing' (50%). Therefore, all models are well suited for relatively highly variable water quality data sets such as those from constructed treatment wetlands (Scholz, 2004).

28.5 Conclusions

This chapter demonstrated the successful application of KNN, SVM and SOM to typical 'black box' systems such as constructed treatment wetlands governed by biochemical processes. The KNN, SVM and SOM methodologies were successfully applied to predict water quality variables of constructed treatment wetlands. The BOD and SS, which are expensive to estimate, can be cost-effectively monitored by applying machine learning tools with input variables such as turbidity and conductivity. Their performances are encouraging, and the use of these models as management tools for the day-to-day process control is therefore recommended.

Moreover, little domain knowledge is required to build a model, and the input variables can be optimized by trial and error.

In terms of accuracy of prediction, the SOM model performed better using smaller input variables in comparison to both KNN and SVM models. Particularly, the SOM model demonstrated its potential to analyze the relationship between complex biochemical variables.

However, establishing the SOM model required more time, if compared to the KNN and SVM models, due to the long searching process of the optimal map. Thus, in addition to the accuracy of model predictions, the resource that is required to build and test the model should be considered when selecting the optimal modeling tool.

In contrast, SUDS systems such as combined attenuation pond and infiltration pond systems (Ellis et al., 2002; EPA, 1999) can be applied as cost-effective local 'source control' drainage solutions; e.g., delaying storm runoff and reducing peak flows. It is often possible to divert all storm runoff for infiltration or storage and subsequent water reuse. As runoff from roads is a major contributor to the quantity of surface water requiring disposal, this is particularly a beneficial approach where suitable ground conditions prevail (Butler and Davies, 2000). Furthermore, infiltration of storm runoff can reduce the concentration of diffuse pollutants such as leaves, feces, metals and hydrocarbons, thereby improving the water quality of surface water runoff (Ellis et al., 2002; Scholz, 2003; Scholz, 2004).

Despite the theoretical benefits of SUDS, the technical constraints associated with the design and operation of large features such as ponds have not been explored due to a lack of experimental data. The rainfall, runoff and infiltration relationships for planted infiltration ponds treating road runoff have not been studied previously. There is also a lack of water quality management data and guidelines for unplanted and planted infiltration ponds that are operated in parallel. Considering the increase in popularity of SUDS, urban planner and developer need to understand the design and operation constraints of systems such as infiltration basins and ponds.

29.2.2 Aim and objectives

The aim is to assess an experimental infiltration pond system (case study) designed according to SUDS guidelines (CIRIA, 2000; EPA, 1999). The objectives are to:

- Identify technical constraints associated with the design and operation of novel planted infiltration ponds;
- Assess the rainfall, runoff and infiltration relationships for unplanted infiltration ponds;
- Assess the water quality, and its management for unplanted and planted infiltration ponds;
- Assess passive and active filamentous green algal control strategies including the use of barley straw and *C. auratus*; and
- Promote the integration of SUDS into urban planning and development.

29.3 Methods

29.3.1 Design of the study site

The pilot plant was designed considering the SUDS guidelines (Building Research Establishment, 1991; Bettes, 1996; CIRIA, 2000; ATV-DVWK-Arbeitsgruppe, 2002) of the British Research Establishment (BRE), Construction Industry

Research and Information Association (CIRIA), and the German Association for Water, Wastewater and Waste (ATV-DVWK). The return period for allowed flooding is ten years. The total road area draining into the SUDS system is approximately 446 m^2. The SUDS is based on a combined wetland and infiltration pond design (Fig. 29.1).

Rainwater runoff from the University road (catchment area) flows directly into a silt trap (Scholz and Zettel, 2004). The silt trap has a maximum capacity of 0.2 m^3. Water from the silt trap overflows via a gravel ditch into the constructed wetland (volume of 2.7 m^3), which also serves the purpose of a below-ground storage tank. If the wetland attenuation system is full, storage water flows over a dry stonewall into a swale and finally into the infiltration ponds.

The ponds can accommodate maximum volumes of 9.7 m^3 each during heavy storm events before flooding of a nearby lawn would occur (Fig. 29.1). The maximum depths of the constructed wetland, and the unplanted and planted infiltration ponds are 0.85, 1.18 and 1.02 m, respectively.

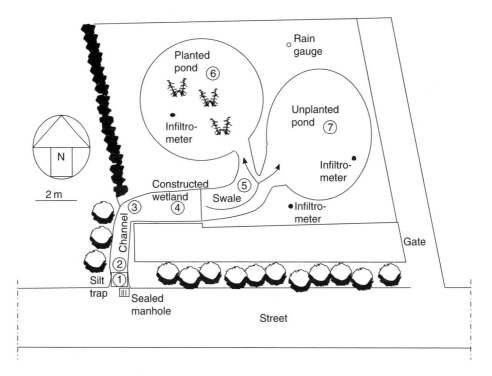

Fig. 29.1 Case study: runoff flows from the road (only eastern part shown) into the silt trap (1), then via the gravel channel (2) into the constructed wetland (3 and 4) and finally via the swale (5) into the infiltration ponds (6 and 7).

In summer 2004, the dominant aquatic plants of the planted infiltration pond were *Typha latifolia* (broadleaf cattail), *Sparganium erectum* (simplestem bur-reed) and *Iris pseudacorus* (paleyellow iris). *Typha latifolia* was planted in the infiltration pond to enhance infiltration. The constructed wetland was planted with *Phragmites australis* (common reed; Scholz et al., 2002), *I. pseudacorus* and *T. latifolia*.

29.3.2 Hydrological methods and water quality analysis

The daily rainfall was monitored by a tipping bucket ARG 100 rain gauge. The infiltration rate of road runoff into the infiltration ponds was determined by using a single-ring infiltrometer. A PN 623-8001 (0.8–2.0 bar) Boart Longyear Interfels total pressure data logger was used to estimate water depth variations in the unplanted pond.

Grab samples at up to seven locations (Fig. 29.1) were predominantly collected from the silt trap (1), constructed wetland (inflow (3) and outflow (4)) and both infiltration ponds (6 and 7). All analytical procedures to determine the water quality were performed according to the USA standard methods (Clesceri et al., 1998).

29.3.3 Fish experiment methodologies

Laboratory experiments with *C. auratus* and different aquatic plants (filamentous green algae, *Elodea canadensis* (Canadian waterweed) and *Callitriche stagnalis* (pond water-starwort)) were carried out in eight aquarium tanks filled up with a mixture of filtered (sieve having a pore diameter of 250 μm) pond water (50%) and tap water (50%): Tank 1: filamentous green algae; Tank 2: *E. canadensis*; Tank 3: *C. stagnalis*; Tank 4: filamentous green algae and *E. canadensis*; Tank 5: *E. canadensis* and *C. stagnalis*; Tank 6: filamentous green algae and *C. stagnalis*; Tanks 7 and 8: filamentous green algae, *E. canadensis* and *C. stagnalis*.

All tanks (except for Tank 8; control) contained six *C. auratus* of similar weight. The overall biomass of plant food per tank was 600 g, and equal proportions of different plants were used. The experiment was stopped after the food sources dropped to <20 g of plant matter.

Twenty healthy *C. auratus* of approximately 180 g total weight were introduced into each infiltration pond on 1 April 2004. The ponds were covered with a plastic mesh to prevent animals such as *Ardea cinerea* (grey heron) and *Felis cattus* (cat) to prey on *C. auratus*.

29.4 Results and discussion

29.4.1 Design and operation of infiltration ponds

The critical storm durations for the BRE, CIRIA and ATV-DVWK design calculations were 0.08, 0.25 and 0.08 h, respectively. The corresponding maximum infiltration pond height requirements were 0.38, 0.15 and 0.12 m, respectively. The estimated mean infiltration rate applied for all methods was 1.17 m/h. A storm duration of 15 min was associated with the maximum pond height and storage volume when CIRIA guidelines were applied (Bettess, 1996; CIRIA, 2000).

All calculated storage volumes (<1 m^3) are considerably lower than 9.7 m^3, so both ponds have more than sufficient storage volumes compared to the theoretical design volumes. It seems that the infiltration rate of 1.17 m/h (SD of 0.21 m/h) is an unrealistically high value for this SUDS system, considering the high spatial variability of the man-made soil at the study site. For example, the infiltration rate at water depths <0.8 m was frequently <0.05 m within 3 d (Fig. 29.2).

The actual design depth for both infiltration ponds (1.10 m) was acceptable, when compared to the BRE (Building Research Establishment, 1991), CIRIA (Bettes, 1996) and ATV-DVWK (ATV-DVWK-Arbeitsgruppe, 2002) guidelines. Signs of system failure have not been observed. However, the strict application of all test guidelines would have led to system failures even during the first few

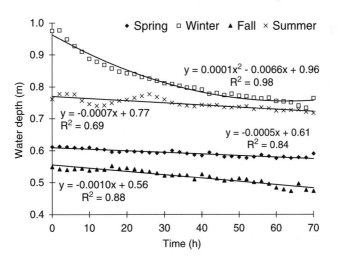

Fig. 29.2 Water-depth changes as a function of the infiltration rates: spring (1 April 2004); summer (28 June 2004); autumn (30 September 2004); winter (28 January 2004).

months of operation. These findings are similar to a previous case study in England discussed elsewhere (Scholz, 2003).

29.4.2 Rainfall, runoff and infiltration relationships

Figure 29.2 gives an indication of representative water-depth variations during periods of virtually no precipitation for all seasons. It can be seen that the infiltration is relatively low at water depths <0.8 m during all seasons except for winter. As a rule of thumb, Eq. (29.4.1) indicates the relationship between infiltration rate and water depth for water depth ranges between 0.3 m and 1 m of the unplanted infiltration pond:

$$\text{water depth (m)} = 100 \text{ (h)} \times \text{infiltration rate (m/h)} + 0.5 \text{ (m)} \qquad (29.4.1)$$

29.4.3 Water quality assessment and management

The water qualities of the constructed wetland inflow, the unplanted pond and the planted pond are shown in Tables 29.1, 29.2 and 29.3, respectively. After one

Table 29.1 Summary statistics: water quality of the inflow to the constructed wetland (1 April 2003 – 30 September 2004)

Variable	Unit	No.	Mean				Standard deviation			
			Time[1]	Time[2]	Time[3]	Time[4]	Time[1]	Time[2]	Time[3]	Time[4]
Temperature	°C	104	11.9	12.3	8.9	14.0	5.5	2.3	4.1	3.8
BOD_5[5]	mg/l	86	17.0	12.3	26.4	16.0	18.2	10.4	25.8	9.8
SS[6]	mg/l	95	385.2	104.4	208.8	601.0	600.3	165.7	305.1	756.9
Ammonia-N	mg/l	67	0.6	0.5	0.5	0.8	0.8	0.4	0.9	0.8
Nitrate-N	mg/l	60	2.5	1.6	1.9	3.2	8.3	2.2	3.4	11.6
Phosphate-P	mg/l	67	0.2	0.0	0.1	0.3	0.2	0.0	0.1	0.2
Conductivity	μS	105	242.2	204.9	276.9	228.9	193.2	174.1	215	180.7
Turbidity	N[8]	106	224.0	133.5	90.5	368.6	442.3	227.0	119.7	605.7
DO[7]	mg/l	100	4.2	3.7	6.0	2.9	6.4	2.7	9.9	1.0
pH	-	106	7.0	7.0	7.0	7.1	0.6	0.2	0.9	0.2

[1] 01/04/2003–30/09/2004.
[2] 01/04/2003–30/09/2003.
[3] 01/10/2003–31/03/2004.
[4] 01/04/2004–30/09/2004.
[5] five-day @ 20°C N-Allylthiourea biochemical oxygen demand (BOD).
[6] suspended solids (SS).
[7] dissolved oxygen (DO).
[8] Nephelometric Turbidity Unit.

**Table 29.2 Summary statistics: water quality of the unplanted pond
(1 April 2003 – 30 September 2004)**

Variable	Unit	No	Mean				Standard deviation			
			Time[1]	Time[2]	Time[3]	Time[4]	Time[1]	Time[2]	Time[3]	Time[4]
Temperature	°C	106	11.8	13.4	6.4	15.6	5.5	2.3	4.1	3.8
BOD$_5$[5]	mg/l	82	19.0	35.9	17.6	14.2	20.4	22.1	25	10.3
SS[6]	mg/l	92	26.4	29.2	24.1	27.1	50.3	24.3	60.1	49.7
Ammonia-N	mg/l	65	0.4	1.4	0.2	0.3	1.2	2.6	0.3	0.7
Nitrate-N	mg/l	63	1.6	0.3	1.3	2.2	6.4	0.1	2.9	9.0
Phosphate-P	mg/l	70	0.3	0.1	0.3	0.3	0.4	0.1	0.5	0.4
Conductivity	μS	107	215.6	246.8	207.1	210.2	121	87.9	156.4	95.2
Turbidity	N[8]	108	17.9	17.3	9.0	25.7	36.9	14.1	19.8	50.6
DO[7]	mg/l	102	5.1	3.8	7.4	3.5	7.9	3.3	12.1	1.2
pH	-	108	7.3	7.0	7.3	7.4	0.4	0.3	0.5	0.2

[1] 01/04/2003–30/09/2004.
[2] 01/04/2003–30/09/2003.
[3] 01/10/2003–31/03/2004.
[4] 01/04/2004–30/09/20004.
[5] five-day @ 20°C N-Allylthiourea biochemical oxygen demand (BOD).
[6] suspended solids (SS).
[7] dissolved oxygen (DO).
[8] Nephelometric Turbidity Unit.

**Table 29.3 Summary statistics: water quality of the planted pond
(1 April 2003 – 30 September 2004)**

Variable	Unit	No	Mean				Standard deviation			
			Time[1]	Time[2]	Time[3]	Time[4]	Time[1]	Time[2]	Time[3]	Time[4]
Temperature	°C	106	11.6	13.2	6.6	15.1	5.5	2.3	4.1	3.8
BOD$_5$[5]	mg/l	88	26.5	26.3	41.8	13.9	32.9	19.4	43.4	17.5
SS[6]	mg/l	95	40.9	38.2	64.7	23.5	87.3	64.9	129.7	34.4
Ammonia-N	mg/l	65	0.2	0.6	0.3	0.1	0.4	1.0	0.3	0.1
Nitrate-N	mg/l	59	0.5	0.5	0.8	0.3	1.7	0.6	2.6	0.8
Phosphate-P	mg/l	67	0.2	0.1	0.2	0.3	0.3	0.1	0.1	0.3
Conductivity	μS	107	288.9	324.7	311.6	254.3	97.4	101.4	124.0	47.7
Turbidity	N[8]	108	17.6	22.6	16.4	16.7	30.6	24.9	16.9	40.5
DO[7]	mg/l	102	4.8	5.8	6.3	3.2	5.3	2.7	7.9	1.3
pH	-	107	7.2	7.1	7.2	7.2	0.3	0.3	0.4	0.1

[1] 01/04/2003–30/09/2004.
[2] 01/04/2003–30/09/2003.
[3] 01/10/2003–31/03/2004.
[4] 01/04/2004–30/09/2004.
[5] five-day @ 20°C N-Allylthiourea biochemical oxygen demand.
[6] suspended solids.
[7] dissolved oxygen.
[8] Nephelometric Turbidity Unit.

year of operation, the water quality of the unplanted infiltration pond (Table 29.2) was acceptable for disposal and recycling according to discussions particularly on biochemical oxygen demand (BOD) and suspended solids (SS) threshold concentrations, elsewhere (Butler and Davies, 2000; Ellis et al., 2002; Scholz, 2003).

Variables indicating the organic strength of both ponds were frequently above international standards (i.e. thresholds of 20 mg/l for BOD and 30 mg/l for SS) for secondary treatment of wastewater (Tchobanoglous et al., 2003). However, water quality monitoring is currently not required for closed systems (zero discharge) in Scotland (CIRIA, 2000).

The BOD and SS concentrations are high in the constructed wetland inflow and both infiltration ponds due to high loads of organic material such as leaves. Removal efficiencies for the SS were >93 and >89% for the unplanted and planted pond, respectively (Tables 29.1 – 29.3). The pH values of the wetland inflow and both ponds are neutral, similar and stable due to algae control measures (see below).

Mats of algae are usually considered unpleasant in their appearance by the public. Therefore, it was necessary to use barley straw bales as a passive algae control method (Ball et al., 2001) for the second summer after system set-up. However, this method does only provide a temporary solution, and does not solve the problem of nutrient accumulation (nitrogen and phosphorus) within the pond sediment.

High temporal data variation indicates the need for at least weekly monitoring of SS to capture concentration peaks exceeding 30 mg/l. In comparison, the BOD concentrations were relatively stable (Scholz, 2003).

Nutrient harvesting was undertaken. The silt trap was regularly emptied and the total wet weight of silt was 55 kg per annum. Aquatic plants of 20 kg were harvested in November 2003. The purpose was to reduce the input of additional nutrients that would otherwise be released from decaying leaves into the planted pond. Furthermore, plant harvesting leads to a prolonged lifetime of the infiltration system by indirectly increasing the storage volume available for sediment during storm events (Scholz, 2003).

29.4.4 Active control of algae with *Carassius auratus*

Carassius auratus is classified as herbivores with wild specimens predominantly feeding on plants. This particularly applies to closed pond systems (Seaman, 1979). Therefore, *C. auratus* could be used to control aquatic weeds (Caquet et al., 1996) and potentially algae in ponds (Richardson et al., 1992). This hypothesis was subsequently tested in the laboratory.

Figure 29.3 indicates that the mean weights and associated SD of *C. auratus* for each aquarium tank appear to be similar at the beginning and end of the

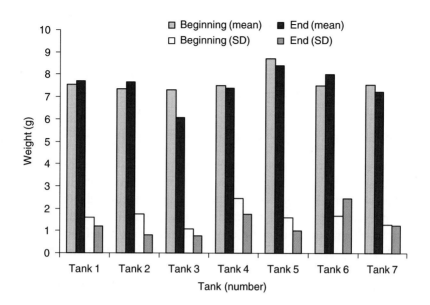

Fig. 29.3 Comparison of means and standard deviations (SD) for weights *of C. auratus* **(common goldfish) at the beginning (9 February 2004) and end (31 March 2004) of the laboratory experiment.**

controlled laboratory experiment. Moreover, an ANOVA showed that all tanks were significantly similar (P < 0.05) concerning their associated fish mean weights, which were measured on eight occasions each.

Carassius auratus seem to feed on all three aquatic plants including filamentous green algae such as *Odeogonium* spp. and *Ulothrix* spp., which were also naturally present in the pond system. However, the greatest drop in fish weights was observed for Tank 3 containing *C. stagnalis* only. Considering that the associated SD was also the lowest, and that *C. stagnalis* belongs to the group of floating aquatic plants that are only accessible by fish below the water level, it can be assumed that *C. auratus* would not have chosen this plant as a preferred food source, if alternative submerged aquatic plants would have been abundant.

Concerning the field experiment, relatively high numbers of filamentous green algae (Chlorophyta) were counted in pond samples taken on 29 March 2004. The dominant alga present was *O. capillare*, which is cosmopolitan in freshwater. *Odeogonium capillare* can form mats in small ponds, and is often mistaken for the more common *Cladophora glomerata* (blanket weed).

Carassius auratus was introduced into both ponds on 1 April 2004 to control filamentous green algae, and to increase public acceptance of SUDS. Concerning algae samples taken on 4 October 2004, both ponds were less dominated by *O. capillare* in comparison to March 2004.

Moreover, the unplanted pond developed a greater diversity of filamentous green algae if compared to the planted pond. This may be due to the absence of macrophytes that would compete with algae for nutrients. Moreover, large macrophytes provide shade leading to a reduction of sunlight penetrating the water, and subsequently reducing the growth of algae.

29.4.5 Integration of SUDS into urban planning and development

The environmental impacts of urban constructions can be reduced with innovative approaches such as 'smart growth' and sustainable urban and green planning (Galuzzi and Pflaum, 1996). Flood protection management and recreational value can be improved by integrating infiltration pond design and operation (in contrast to conventional drainage) into the urban planning and development processes (Campbell and Ogden, 1999; Scholz, 2003). Recreational activities may include watching birds and ornamental fish such as *C. auratus*, walking, fishing, boating, holding picnics and teaching children about aquatic ecology (Galuzzi and Pflaum, 1996).

The confidence of town planners towards SUDS and public acceptance of infiltration ponds can both be increased by correct dimensioning of systems to avoid flooding (Zheng and Baetz, 1999), enhance water pollution control by using a robust pre-treatment train (e.g., silt trap, constructed wetland and swale) and control algae by biological (e.g., *C. auratus*) and not chemical (e.g., copper sulphate) means (Ellis et al., 2002). Moreover, stormwater can be reused for watering gardens and flushing toilets as part of an urban water resources protection strategy.

29.5 Conclusions

This case study described the design and operation of a novel stormwater pond system during the first eighteen months of operation (April 2003 – October 2004). During this period, the SUDS would only have complied with design guidelines if local environmental conditions such as spatial infiltration patterns had been fully considered. Infiltration through the base of both ponds was virtually absent (despite the presence of macrophytes in the planted pond), and should therefore be neglected during the design.

The water qualities of the infiltration ponds were generally unacceptable for water reuse immediately after the set-up period of the SUDS. Biochemical oxygen demand and SS concentrations, in particular, frequently exceeded recognized international secondary wastewater treatment standards, which are, however, not applicable for closed systems with zero discharge. At least weekly water quality monitoring was required to capture temporal data variations.

A bloom of filamentous green algae dominated by *O. capillare* during spring 2004 was observed. Barley straw reduced the growth of algae. Moreover, experiments with *C. auratus* have shown that they eat filamentous green algae equally well as *C. stagnalis* and *E. canadensis*. The presence of tall macrophytes reduced the growth and biodiversity of filamentous green algae.

A successful integration of SUDS into urban regeneration and development can be achieved, if potential design, operation and water quality management problems are addressed during the planning phase. Moreover, current international guidelines for the design and management of infiltration ponds require alterations to avoid system failure as shown in this case study.

Chapter 30

Sustainable urban drainage system model

30.1 Summary

Seventy-nine and 103 sites within Glasgow and Edinburgh, respectively, were identified to assess, if sustainable urban drainage systems (SUDS) can be integrated into future development, regeneration and retrofitting (Chapter 25) plans. A practical SUDS Decision Support Model based on a matrix and weighting system, incorporating the Prevalence Rating Approach for SUDS Techniques (PRAST) has been developed.

The findings indicate that ponds (or lined ponds) and permeable pavements are the most likely individual SUDS techniques, and ponds combined with standard swales (or shallow swales) are the most recommended combination of two SUDS techniques. A separate case-based reasoning (CBR) model compared to the 'linear' SUDS Decision Support Model has also been developed. The output was similar to the 'linear' model.

30.2 Introduction

30.2.1 Sustainable urban drainage systems

Prevention of flooding in urban areas caused by inadequate drainage systems has become a significant problem. With increased development of Greenfield and urban regeneration, potential damage due to prolonged flooding could easily lead to substantial financial losses (Schmitt et al., 2004). The concept of 'source control' for the treatment of stormwater runoff from impermeable surfaces has become widely accepted amongst drainage engineers in both the United States and Europe (Ellis et al., 2004; Villarreal et al., 2004).

Over the past 20 years, the use of best management practice (BMP) in the United States and SUDS in the UK have been instrumental in reducing both the detrimental impact of polluted runoff to the water quality of receiving watercourses, and flooding caused by increased urbanization and traditional stormwater drainage systems. Both BMP and SUDS attempt to mimic the drainage patterns of the natural watershed and can also provide substantial treatment needed to improve the quality of the water discharged to an acceptable standard (Jefferies et al., 1999).

Wetlands generally comprise of a basin with shallow water and aquatic vegetation, which may provide biofiltration. Constructed wetlands remove pollutants through sedimentation, plant uptake, microbial decomposition and filtration. Stormwater wetlands are specifically constructed to treat surface water. They may feature a forebay for suspended solids (SS) capture, and can be designed to allow extended detention during low flows. There is a high proportion of emergent vegetation in relation to open water despite possible benching or stepping of the sides to promote varying water depths. They are not normally designed to provide significant attenuation, but temporary storage may be provided above the permanent water level.

Ponds increase the duration of the flow hydrograph with a consequent reduction in peak flow. Also known as stormwater ponds, retention ponds and wet extended detention ponds, they can be used for attenuation, detention, retention, storage, infiltration and recreational purposes. Wet ponds have a capacity greater than the permanent pond volume, which permits storage of the influent (i.e. stormwater runoff). An overflow outlet to an existing watercourse or sewer is provided for excess flow. These are basins, which have a permanent pool of water in the base for treating incoming stormwater runoff. The primary pollutant removal mechanisms are the settling out of solids and biological degradation activity in the pond. Typically, a residence time of one day gives sufficient treatment as a retention period of one day is not believed to improve the water quality.

Whenever possible, wet ponds should be designed to provide enhanced amenity and wildlife benefits. Performance and maintenance can also be improved by installing a forebay to trap sediments and a wetland around the fringe of the pond perimeter. A catchment of 2 ha is sufficient to maintain the wet pool, which suggests that ponds can be used on most sites. The only restriction is the space required, which limits the use of ponds in constricted city-centre or other urban sites. A liner is required to maintain the wet pool, if the soils below the pond are highly permeable. Alternatively, fully lined ponds can be used, if infiltration is not desired.

An infiltration pond takes runoff from a development, and allows it to infiltrate into the ground. The pond has a sufficient storage volume, so that the runoff may be allowed to empty from the device into the ground over a specific period of time (usually a period of one day to half empty) to provide storage for runoff from any following storms. Infiltration systems reduce the volume of water that has to be disposed off through sewers and provide recharge of groundwater that may maintain water levels in local watercourses. In this manner, they are similar to dry unlined ponds.

An infiltration basin is a dry depression structure designed to promote the infiltration of surface water runoff through the surface of the ground with an emergency spillway (no standard outlet structure exists) to an existing watercourse or sewer. It can be landscaped to give added aesthetic and amenity value. A drawback of infiltration basins is the space required for infiltration, which limits their application in dense urban developments. Infiltration basins should be restricted to a maximum catchment area of 4 ha to reduce the risk and consequences of premature clogging. The removal of pollutants occurs on the infiltration basin's base and also in the underlying soils, where pollutants are filtered and absorbed onto soil particles in the unsaturated zone. In comparison, an infiltration pond is smaller, may permanently contain surface water and is planted.

Standard swales are generally grassed stormwater conveyance channels that use biofiltration and limited infiltration to remove pollutants. When compared to a conventional ditch, a swale channel is broad and shallow and can provide temporary storage for stormwater and reduce peak flows. Swales can form a network within a development scheme, linking SUDS features or conveying water (i.e. predominantly runoff) to a watercourse or sewer. Typically, they are located next to highways, but can also be constructed in landscaped areas within car parks and elsewhere. It is possible to replace conventional piped drainage, and subsequently remove the need for kerbs and gullies.

Shallow swales can be used, if the groundwater table is high and/or there are potential contamination issues with respect to deeper soil layers. However, groundwater should be at least 1 m below the base, and swales should be lined, if contamination is detected and leaching of contaminants is possible. Lined swales should also be used, if infiltration is thought to cause ground instability, foundation problems or even flooding of below-ground building structures.

Swales can be designed for infiltration and pollutant removal, but predominantly transport water from a site. Pollutants can be removed for frequent small storm events, but for less frequent storms (1 in 2 or 1 in 10 year return period), swales act as a storage and conveyance mechanism. It may be difficult to incorporate swales in steeply sloping sites, where it is difficult to retard flow rates. Swales require a relatively slow flow of water over them to remove pollutants and prevent erosion. The speed of flow is dependent on the soil type, but should not exceed 2 m/s for any situation.

Green roofs can reduce the volume and rate of rainfall runoff by using vegetation placed on top of a building to retain water. Below this vegetative surface, there may be soil or substrate, drainage protection, waterproofing and insulation layers.

The extensive green roof covers the entire roof area with low-growing and low maintenance plants. They are accessible only for maintenance. Extensive green roofs typically comprise a 25–125 mm-thick soil layer in which a variety of hardy, drought-tolerant and low plants are grown.

Intensive green roofs are landscaped environments for recreation that include plants or small trees, and are usually publicly accessible. They may also include water features, and storage of rainwater for irrigation. Intensive roofs generally impose far greater loads on the roof structure and require significant on-going maintenance. They may he used over the podium decks to underground car parks.

Infiltration trenches are shallow excavated trenches that have been backfilled with stone to create a below-ground reservoir. Stormwater runoff that is diverted into the trench, gradually infiltrates into the surrounding soil and possibly to the water table. A perforated pipe can be incorporated in the base or a number of perforated pipes at designated depths to promote infiltration.

It is important that infiltration trenches are protected from sediment loads during and after construction until the surrounding runoff area has developed mature ground cover to minimise erosion and sediment transport. The performance of infiltration trenches can be improved, and clogging of the voids between the permeable material reduced by providing pre-treatment in the form of grass filter strips. This will filter particles out of runoff before reaching the infiltration trench. However, such an arrangement may not be practical, if infiltration trenches are located at the sides of highways.

Soakaways are sub-surface structures into which surface water is conveyed. Factors for soakaway success are the size of the area to be drained, the infiltration rate of the soil and the depth to the water table. If the water table is high (i.e. near the surface), a soakaway is not practical. The site of a soakaway must also be located at least 5 m away from any building, and be designed so that it will not saturate the foundations of any structure.

Soakaways take the form of pre-fabricated sections, and are often laid on a simple ring foundation or a strip footing of concrete that supports the weight of the chamber section. A cover is essential to allow access for cleaning and inspection. A soakaway should not require any maintenance, but it is good practice to check the structure every few months for silting or contamination. If significant silt build-up occurs on a regular basis, it may be worthwhile constructing a 'catch pit' to intercept silt from the inlet. Any build-up of silt at the base of the soakaway can be removed manually during dry conditions when the soakaway is empty.

Filter strips are vegetated strips of land that act as buffers by accepting storm runoff from adjacent impermeable areas. The runoff is designed to sheet flow across the filter strip sufficiently slowly to filter out sediment and associated pollutants. The main restriction to the use of filter strips is the large space that they require. They are most suited to small car parks, roads and similar areas, but are not generally suitable for use in dense urban developments. If sheet flow changes to concentrated flow, as is possible over impervious surfaces, then the velocity is too great for filtration to be effective, thus filter strips should be used to treat only very small drainage areas. Filter strips have treatment mechanisms similar to that of swales, but runoff should be evenly distributed across the surface of these structures.

Grassed filter strips behave in a similar way to grassed channels, and many of the design methods and criteria for swales can also be applied to filter strips. They can be used in most ground conditions, and must be constructed to provide an even and consistent slope with no depressions that will cause localised ponding or promote flow in channels.

Pervious surfaces are pavement constructions that allow rainwater to infiltrate through the surface and into the underlying construction layers. Water can be stored prior to infiltration to the ground, reused or released to a surface watercourse or other drainage system. Pervious surfaces can be either porous or permeable.

Permeable pavements can be used to reduce runoff rates and volumes from urban areas. They attenuate the runoff while preserving the value of the area for urban development. The surface infiltration rate should be greater than the design rainfall intensity and include any runoff from adjacent impermeable areas. The storage volume can be calculated based on the volume and porosity of the underlying storage layer. The storage volume should take between one and two days to empty.

A below-ground storage structure should be designed to accumulate surface water runoff, and to release it after a lag period to prolong the runoff hydrograph and to decrease the peak flow. The structure may contain aggregates or plastic detention cells (e.g., Atlantis Water Management Ltd. system) and act also as a water recycler or infiltration device. Subsurface storage means that pre-development runoff levels can be sustained without constructing ponds that use valuable space and create potential health and safety hazards.

Below-ground stormwater storage means that above-ground space can be put to alternative use. Depending on the site, the space saved could be used for parking space, more area for the construction of buildings or enhanced landscaping. This SUDS option is particularly useful for dense city centres or retail areas, where the space for other SUDS options is limited.

Below-ground storage systems can be successfully combined with pervious or permeable surfaces to provide a SUDS treatment train. Various permeable pavement solutions are on the market; e.g., systems provided by Formpave Ltd.

A water playground is a small pond-like feature integrating play facilities for children and acting as an amenity feature to increase public acceptance with little impact to surface water runoff attenuation. Water playgrounds are environments in which people interact with water and each other. Children can participate by turning valves, pulling levers and controlling the direction and flow of water in a park or other recreational setting.

Water playgrounds are ideal additions to city and residential parks, and are also well suited to commercial and retail installations. Properly designed and maintained water playgrounds are as safe as commercial swimming pools, and should only be installed when the water quality and cleanliness is of a suitable standard to be handled by humans.

30.2.2 SUDS impact on water quantity and quality

Rapid urbanization and its consequent increase in impermeable surface areas and changes in land use generally result in problems of flooding and heavy pollution of urban streams and other receiving watercourses. This is frequently coupled with groundwater depletion and a threat to natural water resources (Andoh and Declerck, 1997).

The traditional drainage system is no longer adequate in many areas, especially in high density housing estates. The subsequent effects can be life threatening, and may lead to damage to buildings and other infrastructure assets, and disrupt the functioning of business and society (Mark et al., 2004; Schmitt et al., 2004). Through SUDS techniques, water is either infiltrated or conveyed more slowly to watercourses or sewer systems via ponds, swales, filter strips or other sustainable installations, which could reduce the peak flow during heavy storms (Pettersson, 1997).

Stormwater in open systems forms the basis for recreation and the development of ecosystems with a diverse animal and plant life. In many ways, the principles of alternative stormwater management adopt the natural processes in the environment, and adapt them to urban conditions and requirements. At the same time, the effects of pollution are taken into account (Astebøla et al., 2004).

Butler and Parkinson (1997) mentioned that SUDS should maintain a good public health barrier, and avoid local or distant pollution of the environment. Non-point sources of pollution are difficult to identify and control, and are one of the main reasons that urban rivers fail to reach the water quality objectives set for them, while SUDS techniques are available to help combat diffuse pollution (Mitchell, 2005). For instance, the runoff from roofs and streets contribute between 50 and 80% of heavy metals to the total mass flow in domestic sewage (Butler and Parkinson, 1997).

30.2.3 Development and regeneration in Glasgow

Glasgow is associated with many areas for which the need for SUDS is becoming apparent. In 2003, the new 'City Plan' was adopted, outlining a large number of urban areas known as 'New Neighbourhoods' in which redevelopment should occur. The main urban drainage concern for Scottish Water and Glasgow City Council is the lack of sewer system capacities for additional surface runoff.

With future development and regeneration activities, this situation is likely to become worse, and the need for SUDS implementation becomes vital to

Glasgow's continuing expansion. New guidelines incorporating a decision support tool that identifies SUDS techniques, which are feasible for a particular site over a range of boundary and environmental conditions, are required.

Furthermore, a previous project undertaken by Hyder Consulting for the East End of Glasgow did not demonstrate how SUDS implementation could lead to a significant reduction of potential combined sewer overflow spills. However, the desk study failed to incorporate actual site conditions such as the percentage of build up areas or the local topography of the landscape including watercourses and hills.

30.2.4 Sustainable drainage systems in Edinburgh

Construction projects in Edinburgh's outskirts are dominated by Greenfield development activities. City regeneration projects are predominantly restricted to Brownfield sites. Current and proposed building projects within the city boundary are the City Centre, the Granton Waterfront, Edinburgh Park, South Gyle and Sighthill. These areas are likely to become commercial or high density residential developments. Within the City, there are designated recreational areas, which can be used to drain runoff after SUDS implementation. This process of utilising recreational space is a less controversial form of SUDS retrofitting.

As development and regeneration activities increase, the amount of impermeable surfaces also increases. It follows that there is a pressing need to use SUDS techniques to control runoff by ground infiltration or storage, which has beneficial impacts on downstream catchments (City of Edinburgh Council, 1999).

Flooding within the City of Edinburgh is mostly caused by rainfall in the upper river catchment areas, which lie outside the boundaries of the city. Recent flooding has occurred predominantly in the following catchments and localities: River Almond and Gogar Burn; Ferry Burn; South Queensferry; Linn Mill Burn; Braid Burn and Figgate Burn; Burdiehouse, Niddrie and Brunstane Burn (City of Edinburgh Council, 1999).

30.2.5 Aims and objectives

The main aims are to establish new SUDS decision support tools for planners, and to compare the outcomes for Edinburgh and Glasgow with each other. The detailed objectives include the following:

- To identify suitable SUDS sites within Glasgow and Edinburgh;
- To classify qualitatively and quantitatively sites suitable for different SUDS technologies;
- To outline a SUDS Decision Support Model for development, regeneration and retrofitting sites;
- To publish the model on the website, and to evaluate feedback from users;

Table 30.1 Data base summary for Glasgow (79 sites) and Edinburgh (103 sites) representing the current situation but taking into account already planned construction work

Contamination	Yes: 9/0	No: 91/100			
Possible SUDS area (%)	≤20: 1/8	20< x ≤60: 23/4	60< x ≤80: 23 / 1	>80: 53/87	
Land values	Low: 19/33	Low-medium: 30/19	Medium: 39/35	Medium-high: 9/9	High: 3/4
Runoff quantity	Low: 27/46	Low-medium: 24/22	Medium: 42/25	Medium-high: 5/6	High: 3/1
Runoff quality	Poor: 5/0	Average: 52/39	Good: 43/61		
Roof runoff (%)	≤50: 63/6	>50: 37/94			
Car park runoff (%)	≤50: 85/88	>50: 15/12			
Road runoff (%)	≤50: 65/88	>50: 35/12			
Roads[a]	Motorways: 9/0	Primary road: 15/12	A road: 27/18	B road: 14/3	Others: 35/67
Drainage type	Sewer only: 60/58	Watercourse: 39/40	Not currently detected: 1/2		
Groundwater	High: 22/15	Low: 62/68	Currently not determined: 16/17		
Soil infiltration	High: 62/68	Low: 22/15	Currently not determined: 16/17		
Impermeable surface (%)	≤20: 3/53	20< x ≤40: 52/14	40< x ≤60: 16/7	60< x ≤80: 19/8	>80: 10/18
Catchment size (m²)	≤50000: 30/46	50000< x ≤100000: 28/30	100000< x ≤200000: 32/13	x > 200000: 10/11	
Slope (x in 100 m)	≤1: 19/21	1< x ≤5: 49/ 53	5< x ≤10: 16/19	x>10: 16/7	
Ownership	Council: 62/39	Private: 19/35	Council and private: 19/26		
Ecological impact	Yes: 24/26	No: 76/74			
Acceptance warning	Green flag: 82/81	Orange flag: 18/17	Red flag: 0/2		
Site classification	Development: 35/21	Regeneration: 35/38	Retrofitting: 11/17	Retrofit with parks: 19/24	

[a]Higher classified roads take precedence over any lower classified roads, and score the associated bin entry.
Note: Proportions are expressed in %.

Table 30.2 Dominant (d) and supplementary (s) sustainable urban drainage system (SUDS) decision support criteria

SUDS technique	Catchment size (m²)	Area for SUDS (m²)	Future runoff (1–5)	Runoff quality	Land contamination	Land value	Ownership fragmented	Groundwater level
Wetland	>50 000 (s)	>5000 (d)	>2 (d)	Average (d)	No (d)	<4 (d)	No (s)	–
Pond	>15 000 (s)	>50 (d)	–	Average (d)	No (d)	<4 (s)	–	–
Lined pond	>15 000 (s)	>50 (d)	–	–	–	<4 (s)	–	–
Infiltration basin	>15 000 (s)	>50 (d)	<4 (s)	Average (d)	No (d)	<4 (s)	–	Low (d)
Standard swale	–	>200 (s)	<4 (s)	Average (s)	No (d)	<4 (s)	No (s)	Low (s)
Shallow swale	–	>200 (s)	<3 (s)	Average (s)	No (s)	<4 (s)	No (s)	–
Filter strip	>15 000 (s)	>500 (d)	<4 (s)	Average (d)	No (d)	<4 (d)	No (d)	Low (d)
Soakaway	>3000 (s)	>200 (s)	<3 (s)	Average (s)	No (d)	<5 (s)	–	Low (d)
Infiltration trench	>3000 (s)	>50 (s)	<4 (d)	Average (s)	No (d)	–	No (s)	Low (d)
Permeable pavement	–	–	<4 (s)	Average (s)	No (d)	–	–	–
Below-ground storage	>5000 (s)	>50 (d)	<5 (s)	–	–	–	–	Low (d)
Water playground	>3000 (s)	>10 (s)	<3 (s)	Good (d)	No (d)	–	–	–
Green roof	–	>20 (s)	<4 (s)	Good (s)	–	–	No (s)	–
Swale + pond	>20 000 (s)	>300 (d)	–	Average (d)	No (d)	<4 (s)	No (s)	Low (s)
Shallow swale + pond	>20 000 (s)	>250 (d)	<4 (s)	Average (d)	No (d)	<4 (s)	No (s)	–
Infiltration trench + Below-ground storage	>8000 (s)	>150 (d)	<5 (s)	Average (s)	No (d)	–	No (s)	Low (s)

Table 30.2 – Cont'd

SUDS technique	Slope (x m in 100 m)	Maximum slope	Soil infiltration	Ecological impact	Impermeable area < (%)	Impermeable area > (%)	Drainage to watercourse or sewer
Wetland	–	15 (s)	–	Yes (s)	40 (s)	–	Yes (d)
Pond	–	20 (s)	–	–	65 (s)	–	Yes (d)
Lined pond	–	20 (s)	–	–	65 (s)	–	Yes (d)
Infiltration basin	–	30 (s)	High (d)	–	60 (s)	–	–
Standard swale	>1 (s)	10 (s)	–	–	85 (s)	–	Yes (s)
Shallow swale	>1 (s)	15 (s)	–	–	90 (s)	–	–
Filter strip	>2 (s)	40 (s)	High (s)	–	50 (s)	–	–
Soakaway	>1 (s)	25 (s)	High (d)	–	90 (s)	–	–
Infiltration trench	>1 (s)	15 (s)	High (d)	–	–	30 (s)	–
Permeable pavement	–	20 (s)	–	–	–	30 (s)	–
Below-ground storage	–	15 (s)	–	–	–	40 (s)	Yes (d)
Water playground	–	20 (s)	–	–	–	–	–
Green roof	–	20 (s)	–	–	–	–	Yes (d)
Swale + pond	>1 (s)	10 (s)	–	–	55 (s)	–	Yes (d)
Shallow swale + pond	>1 (s)	15 (s)	–	–	60 (s)	–	Yes (d)
Infiltration trench +	>1 (s)	15 (s)	–	–	–	40 (s)	Yes (d)
Below-ground storage							

Table 30.3 Weightings for the corresponding supplementary SUDS decision support criteria

SUDS technique	Catchment size (m²)	Area for SUDS (m²)	Future runoff (1-5)	Runoff quality	Land contamination	Land value	Ownership fragmented	Ground water level
Wetland	2	0	0	0	0	0	2	0
Pond	1	0	0	0	0	2	0	0
Lined pond	1	0	0	0	0	2	0	0
Infiltration basin	2	0	2	0	0	2	0	0
Standard swale	0	1	3	2	0	1	3	2
Shallow swale	0	1	2	2	1	1	2	0
Filter strip	2	0	3	0	0	0	0	0
Soakaway	2	2	3	2	0	2	2	0
Infiltration trench	1	1	0	2	0	0	2	0
Permeable pavement	0	0	2	2	0	0	0	0
Below-ground storage	1	0	2	0	0	0	0	0
Water playground	1	1	2	0	0	0	0	0
Green roof	0	1	3	2	0	0	2	0
Swale + pond	2	0	0	0	0	2	2	2
Shallow swale + pond	2	0	3	0	0	2	2	0
Infiltration trench + below-ground storage	1	0	3	3	0	0	2	2

of Britain and the USA in the future. For the Glasgow and Edinburgh SUDS Management Project, 182 sites were available to build up a database of 'cases', which has a high variety of different sites. A suitable SUDS technique for a new site is suggested based on a comparison between the new site information and previous 'cases'. The recommended SUDS option and the PRAST output can be obtained through both the 'linear' and the CBR models.

30.4 Results and discussion

30.4.1 SUDS Decision Support Model output

The SUDS Decision Support Model output is based on the raw data of 182 sites (Table 30.1 and Fig. 30.1). Sites in Edinburgh have more roof runoff in comparison to sites in Glasgow, because of the relatively higher housing density. The proportion of properties owned by the Council is higher in Glasgow, if compared to Edinburgh, reflecting the high proportion of the population close to the poverty line. Glasgow has considerably more regeneration sites compared to Edinburgh that mainly relies on SUDS retrofitting due to a lack of affordable open space.

The SUDS variables 'land value' and 'runoff quantity' are influenced greatly by property purchasing costs and available storage volumes, respectively. They are the most influential variables for both cities. 'Land value' is estimated as a relative variable. The high runoff volumes in Glasgow reflect higher rainfall depths and larger catchment areas in comparison to Edinburgh (Table 30.1).

Table 30.5 shows the proportions of four SUDS Decision Support Model categories (i.e. non applicable, applicable, recommended option and best option) for different SUDS techniques applied for sites in Glasgow (G) and Edinburgh (E). Ponds (G: 59%; E: 60%), lined ponds (G: 65%; E: 61%) and permeable pavements (G: 57%; E: 71%) obtained high proportions for best SUDS options considering individual SUDS techniques for both Glasgow and Edinburgh. Moreover, ponds combined with swales (G: 19%; E: 29%) or shallow standard swales (G: 35%; E: 38%) are the most likely choices for SUDS combinations. This output is reasonable considering general technical judgement.

Ponds (or lined ponds) increase the duration of the flow hydrograph with the consequence of a reduction in the peak flow, which is considered to be the most effective SUDS technique to control stormwater quantity and quality (Pettersson, 1997). Permeable pavements require little space and could be applicable in most situations. Standard and shallow swales are generally grassed stormwater conveyance channels that use bio-filtration and limited ground infiltration to remove pollutants. Swales can also form a network within a SUDS development scheme, linking SUDS techniques and conveying runoff to a watercourse or sewer.

Table 30.5 Proportions of weightings for the linear SUDS Decision Support Model and CBR model applied for Glasgow and Edinburgh (%)

Glasgow / Edinburgh sites (%)	SUDS Decision Support Model				CBR model			
	-	x	xx	xxx	-	x	xx	xxx
Wetland	80	8	3	10	82	8	4	6
Pond	13	9	19	59	10	16	30	43
Lined pond	1	11	23	65	3	16	34	47
Infiltration basin	34	19	11	35	33	15	23	29
Standard swale	8	10	61	22	6	11	48	34
Shallow swale	0	25	53	22	0	25	51	24
Filter strip	54	10	10	25	49	16	19	15
Soakaway	29	10	33	28	28	35	6	30
Infiltration trench	37	10	30	23	35	11	39	14
Permeable pavement	8	13	23	57	5	15	42	38
Below-ground storage	23	14	30	33	22	20	30	28
Water playground	58	10	4	28	61	11	5	23
Green roof	4	14	47	35	6	48	15	30
Swale + pond	13	24	44	19	13	29	34	24
Shallow swale + pond	13	16	35	35	11	19	34	35
Infiltration trench + below-ground storage	9	29	54	8	5	29	57	9
Wetland	85	9	4	2	88	7	3	2
Pond	5	11	24	60	2	16	39	44
Lined pond	4	11	24	61	2	15	39	45
Infiltration ba sin	16	28	9	48	15	19	30	36
Standard swale	1	13	38	49	0	10	37	53
Shallow swale	0	15	45	41	0	12	47	42
Filter strip	32	17	24	27	29	20	36	15
Soakaway	16	8	35	42	15	23	17	46
Infiltration trench	21	5	39	35	19	4	51	25
Permeable pavement	1	8	20	71	0	6	39	55
Below-ground storage	17	7	48	29	15	14	53	18
Water playground	40	14	7	40	44	12	15	30
Green roof	4	13	34	50	8	32	16	45
Swale + pond	7	30	34	29	4	30	44	22
Shallow swale + pond	7	31	24	38	2	25	45	28
Infiltration trench + below-ground storage	6	15	62	17	2	16	64	18

'-'   = non applicable.
'x'   = applicable.
'xx' = recommended option.
'xxx' = best option.

Comparing the output from Glasgow and Edinburgh with each other, Edinburgh sites are more suitable for below-ground SUDS techniques such as infiltration trenches, permeable pavements and below-ground storage facilities. Below-ground storage is more suitable for retrofitting of sites in Edinburgh due to a lack of affordable open space. For both cities, the wetlands option obtained a high proportion of 'non applicable' entries although wetlands are considered to be the most ecological SUDS technique by environmentalists. However, traditional constructed wetlands usually require relatively large construction areas, and high ecological impact potentials.

The 14 demonstration sites in Glasgow and Edinburgh were selected based on the rationale that demonstration sites should represent different geographical areas, types of land use, site classification types, and SUDS techniques to be implemented. Moreover, the detailed design of demonstration sites would help engineers and planners to have a better understanding of SUDS.

A free trial version of the SUDS Decision Support Model has been published on the internet (http://www.see.ed.ac.uk/research/IIE/research/environ/uw12.html). Users are encouraged to provide the author with feedback.

30.4.2 PRAST analysis

As outlined in the methodology section, both individual and combinations of SUDS techniques were rated on the PRAST scale (Table 30.4). Rating methods based on civil engineering and sustainability perceptions were explored. With respect to Glasgow and Edinburgh, civil engineers are predominantly concerned with reducing the peak flow to alleviate the risk of flooding further downstream, and minimizing the space for SUDS implementation to utilize the majority of available land for construction. Correspondingly, environmentalists are predominantly concerned with increasing the ecological impact of SUDS to enhance biodiversity.

For the purpose of this modeling example, civil engineering rating obtained 4 out of 10, and sustainability rating 6 out of 10 weighting points. The final SUDS technique positioning after PRAST implementation (example only) is shown in Table 30.6. Findings show that combinations of ponds and swales obtained the highest score (see above). In comparison, the approach rates infiltration trenches as the least favourable SUDS technique, because of the frequently insufficient engineering performance and low environmental benefit (Jefferies et al., 1999).

The PRAST scale was applied to all SUDS possibilities, and recommended the highest ranked SUDS techniques as the solution. The numbers and corresponding proportions of individual and dual SUDS techniques derived from the PRAST analysis are shown in Table 30.6. Noticeable, the combination of standard swales (or shallow swales) with ponds and ponds alone have the greatest proportion of recommended SUDS solutions for both sites in Glasgow and Edinburgh. Wetlands, filter strips and some other SUDS features did not score at all. However, these techniques were potentially suitable for some sites, but a different SUDS technique, which featured higher on the PRAST scale, was more suitable.

Table 30.6 Results of the Sustainable Urban Drainage System (SUDS) Decision Support Model and case-based reasoning model for the implementation of SUDS techniques using the PRAST scale (Table 30.4) for both Glasgow and Edinburgh sites

PRAST position	SUDS techniques	Glasgow sites/Edinburgh sites			
		SUDS Decision Support Model		CBR model	
		Number of sites	% of sites (%)	Number of sites	% of sites (%)
1	Standard swale + ponds	15/30	19/29	19/23	24/22
2	Shallow swales + ponds	14/9	18/9	11/6	14/6
3	Attenuation/detention ponds	18/23	23/22	5/17	6/17
4	Wetlands	0/0	0/0	0/0	0/0
5	Lined attenuation/detention ponds	4/1	5/1	3/1	4/1
6	Infiltration basin/ponds	1/1	1/1	0/2	0/2
7	Standard swales	3/12	4/12	9/22	11/21
8	Shallow swales	1/0	1/0	2/2	3/2
9	Infiltration trenches + below-ground storage	1/3	1/3	4/0	5/0
10	Green roof	5/9	6/9	2/9	3/9
11	Permeable pavement	10/13	13/13	11/16	14/16
12	Filter strips	0/0	0/0	0/0	0/0
13	Soakaways	0/0	0/0	0/1	0/1
14	Supplementary water playground	0/0	0/0	0/0	0/0
15	Below-ground storage	6/0	8/0	7/1	9/1
16	Infiltration trenches	0/0	0/0	0/0	0/0
17	No SUDS possible	1/2	1/2	6/3	8/3

Moreover, there is one site in Glasgow and there are two sites in Edinburgh that are not associated with any recommended SUDS technique, mainly because of steep slopes on these sites (Table 30.6).

30.4.3 Case-based reasoning model output

The output of the SUDS Decision Support Model incorporating PRAST is shown in Tables 30.5 and 30.6. The output from the CBR model is similar to the SUDS Decision Support Model output. The novel idea is to use CBR in the decision-making process for SUDS installations. However, a larger 'case base' would have been an advantage in order to increase the relevance of this technique.

30.5 Conclusions

A survey of 182 sites in Glasgow and Edinburgh indicated that it is feasible to implement different SUDS techniques and short SUDS treatment trains within both cities, which are fundamentally different from each other. A general SUDS implementation matrix (including dominant and supplementary criteria), which can be adapted to other cities and countries has been outlined.

A practical SUDS Decision Support Model based on a support matrix (SUDS variable specifications for SUDS techniques) and an associated weighting system has been developed to give the practitioner a novel tool to assess the suitability of different SUDS techniques for a particular site with and without applying his or her own judgement. A trial SUDS Decision Support Model version has been published on The University of Edinburgh's website to obtain user feedback.

The modeling outcome indicates that ponds (or lined ponds) and permeable pavement are the most frequently proposed SUDS techniques for Glasgow and Edinburgh, and ponds combined with standard or shallow swales are the most recommended SUDS combinations. Fourteen SUDS demonstration areas, which are representative for both different sustainable drainage techniques and geographical areas available for development, regeneration or retrofitting within Glasgow and Edinburgh, have been identified.

The PRAST, which allows for the application of different proportions of civil engineering and sustainability judgement used in SUDS decision-making to be made, was developed. Ponds linked to swales (or shallow swales) were placed at the top (most suitable) after PRAST ranking was undertaken. They were also associated with the highest proportion of recommended SUDS solutions for both Glasgow and Edinburgh.

A separate CBR model was compared to the SUDS Decision Support Model. The outputs of both models were similar indicating that the linear model works similar to the human mind.

Chapter 31

Natural wetlands treating diffuse pollution

31.1 Case study overview

31.1.1 Summary

The Environment Ministry of Schleswig-Holstein (Northern Germany) has implemented a novel peatland rehabilitation programme to utilise the high denitrification potential of degenerated and minerotrophic peatlands for the reduction of nitrate input into aquatic ecosystems, and to simultaneously improve the habitat conditions. The realisation of both goals required changes to the water management strategy to adapt it to the site-specific geohydrological conditions. The effect of raised groundwater levels and extensive land use on the water quality of heavily vegetated and groundwater-fed ditches was assessed in a riparian peatland located in the River Eider Valley (an internationally important wetland case study area).

The water quality of the selected representative ditches was regularly assessed at different discharge levels during different flow obstruction cover periods. The hydraulic residence time was predominantly a function of the ditch geometry and the overall flow obstruction. A better understanding of the effect of ditch vegetation on the temporal flow patterns and the hydraulic residence times is of high environmental interest, especially for improving nutrient standards in lowland rivers such as the River Eider. Within-ditch vegetation and other hydraulic flow obstructions such as accumulated silt and organic debris increased the hydraulic residence time and lead to an improvement of the water quality (e.g., reduction in nitrate content) along the ditch.

While ortho-phosphate and ammonia concentrations were acceptable to German water quality standards, nitrate-nitrogen concentrations were frequently elevated due to high discharges despite of high flow obstruction predominantly due to macrophytes. Further findings show that the lower stretches of the ditches were flooded by the River Eider due to the absence of a macrophyte-mowing scheme that lead to increased water levels during late summer.

31.1.2 Overview of the content

This section highlights the wider case study context as well as water management and nutrient control issues. The fundamentals of nutrient transformations and removal processes are also explained. This section is followed by a detailed explanation of the aims and objectives of the field-scale case study. Information on ditch hydraulics, water quality and vegetation is required to assess the impact of water management techniques on the functioning of ditch ecosystems within a wider catchment.

The sections on ditches and the channel of the River Eider within the River Eider Valley reveal details about the case study and water sampling. These Sections are followed by method descriptions on watercourse classification within the River Eider Valley, discharge determination for open channels, water quality analyses, vegetation characterisation and data analysis.

Findings with respect to the characteristics of watercourses in the River Eider Valley, and their corresponding vegetation characteristics and water quality during spring and summer are presented. The disappearence of ditches due to unhindered vegetation growth is discussed. Specific reference to the interpretation of both the water quality variations within the study area and vegetation characteristics for selected ditches is made.

31.2 Introduction

31.2.1 Background of the case study

Phosphorus concentrations in most freshwater bodies in Germany showed a decreasing trend after the implementation of advanced technical solutions in sewage treatment plants (e.g., phosphate stripping). These measures only slightly reduced total nitrogen concentrations. As a consequence, their concentrations have altered little since 1980 (European Environment Agency, 1999; Umweltbundesamt, 2001).

Presently, agriculture is frequently the main source of diffuse nitrogen and phosphorus pollution inputs to water bodies in Western Europe (European Environment Agency, 1999). A further reduction of nitrogen concentrations in freshwater bodies requires interventions during all stages of nitrogen flow through the landscape. Nitrogen loss from unsaturated soil layers to the saturated soil layers can be reduced by the adoption of best land use practice (Frede and Dabbert, 1998; Reiche, 1994). A reduction of nitrate occurs during transport through groundwater, peatlands and wetlands (Arheimer and Brandt, 1998). Burt (2003) has reviewed the monitoring of nitrate for selected hydrological systems.

Quantifying nutrient retention in natural and constructed wetlands, and in constructed treatment reed beds, is an important scientific and environmental task when enhancing water quality (Mitsch and Gosselink, 2000; Scholz, 2003; Scholz and Xu, 2002; Vymazal et al., 1998). In general, the abiotic and biotic conditions in wetlands support several biogeochemical and physical processes resulting in reduced nutrient concentrations in the outflowing water compared to the inflowing water. Therefore, wetland and river restoration are frequently suggested as an effective measure to combat eutrophication of aquatic ecosystems and to reduce nutrient loads to the sea.

In groundwater discharge areas such as the River Eider Valley, groundwater buffer zones are an effective means for reducing the total nitrogen concentration (predominantly nitrate-nitrogen) before the polluted load reaches the River Eider, predominantly via small ditches (Trepel and Kluge, 2002). Rivers themselves support nitrogen removal processes depending, for example, on macrophyte composition and stream morphology (Eriksson and Weisner, 1999).

31.2.2 Nutrient transformations and removal processes

A further reduction of nutrient input into aquatic water bodies can be achieved by two strategies: a reduction of nutrient losses by the adoption of a best land use practice, and by using natural nutrient removal and transformation processes. A successful adoption of the latter strategy requires a basic understanding of the involved processes.

In the nitrogen cycle, ammonification, nitrification and denitrification are the three dominant transformation processes. Ammonification is the conversion from organic-nitrogen to ammonia-nitrogen. Ammonia-nitrogen is used by the aquatic plants and micro-organisms for new biomass development. Ammonification is slower in anaerobic than in aerobic soils, because of the reduced efficiency of heterotrophic decomposition in anaerobic environments. Ammonification also depends on the pH being within an optimum range of approximately 6.5–8.5.

Furthermore, nitrification transforms ammonia-nitrogen to nitrate-nitrogen. This transformation has two steps. Ammonia-nitrogen originates from decomposed plants and animals. Nitrification (i.e. the conversion of ammonium-nitrogen to nitrate-nitrogen) is important, because plants often take up nitrate-nitrogen preferentially to ammonia-nitrogen. However, this transformation requires oxygen.

Denitrification is the process in which nitrate-nitrogen is reduced to gaseous nitrogen. This transformation is supported by facultative anaerobes. These organisms are capable of breaking down oxygen-containing compounds such as nitrate-nitrogen to obtain oxygen in an anoxic environment. For a more detailed discussion on the nitrogen cycle including nitrogen transformations, the reader is referred to Nuttall et al. (1997) and Davidson et al. (2002).

31.2.3 Aim and objectives

The aim of this chapter is to assess the temporal and spatial hydraulic and bio-chemical functioning of groundwater-fed ditch ecosystems and their contribution to the overall catchment and particularly nutrient dynamics of a riparian peatland located in the River Eider Valley. The data gathered should help society to understand the challenges of holistic wetland system management, diffuse pollution, and the linking scales in catchment management. The objectives are as follows:

- To characterise all watercourses and classify ditches at a particular stretch of the River Eider;
- To determine ditch discharges by carrying out experiments with temporary weirs at different hydraulic loading rates and different flow obstruction covers in ditches;
- To assess the water quality of the selected ditches and the River Eider between May and September;
- To describe the effect of summer flooding on the overall water quality of the buffer zone;
- To characterise the vegetation at the case study site, particularly, at those areas close to the ditches;
- To assess the effect of macrophytes and channel morphologies (river and ditches) on nutrient retention; and
- To assess if reduced ditch maintenance is an option to achieve nutrient reduction.

31.3 Materials and methods

31.3.1 Case study and sampling

The River Eider Valley is a riparian freshwater wetland of 257 ha in Northern Germany ($10°3'$ East and $54°11'$ to $54°15'$ North; Fig. 31.1). The River Eider Valley is the name of a specific site located between the villages Schmalstede (upstream) and Techelsdorf near Flintbek (downstream), and not the name for the valley of the River Eider as a whole. This valley is located 10 km south of Kiel (capital of the federal state of Schleswig-Holstein). The climate is usually humid and cool-temperate, with an annual mean temperature of approximately 8°C and a mean annual precipitation of approximately 800 mm/year (Trepel et al., 2003).

The River Eider Valley peatland is hydrologically characterised by groundwater inflow from the surrounding hills and river water inflow from the upstream catchment basin with an area of 120 km^2 (van der Aa et al., 2001; Jensen et al., 2001).

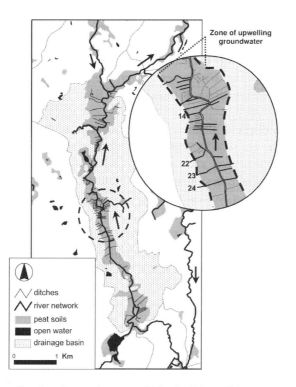

Fig. 31.1 Overview showing the study area within the River Eider Valley, the River Eider, the main ditches with measurable discharge since June 2003, and the selected representative Ditches 14, 22, 23 and 24. The zones of upwelling groundwater on each side of the river are indicated by broken lines.

Mire genesis was dominated first by river water inflow from the upstream catchment area, and in a later successional stage by groundwater inflow at the mire margins (Trepel and Kluge, 2002).

Active drainage with ditches and tile drains, land use intensification, cutting of turf by hand (Klove, 2001) and river regulation have resulted in severe degeneration of the peatland. Altering of the water flow patterns led to the mobilisation of nutrients from decaying peat vegetation, and resulted in an increased nutrient load for most ditches and some river stretches (Bockholt et al., 1992; Trepel et al., 2003).

31.3.2 Ditches and the channel of the River Eider

The study area is a specific site within the River Eider Valley. Figure 31.1 gives an overview of the study area showing the River Eider, the main ditches and the locations of the studied selected representative Ditches 14, 22, 23 and 24. The River Eider has a discharge range between 0.3 and 4 m^3/s (Trepel et al., 2003).

Table 31.1 Water quality comparison between small watercourses and the River Eider

Variable	Statistics	Schmalstede (upstream)	Watercourses (in the valley)	Flintbek (downstream)
Water temperature	Mean	13.8	12.0	13.6
(°C)	Standard deviation	1.68	2.39	1.86
	Count	17	72	17
pH	Mean	7.9	7.5	7.7
(-)	Standard deviation	0.10	0.26	0.12
	Count	17	56	17
Nitrate-nitrogen*	Mean	3.1	1.5	2.2
(mg/l)	Standard deviation	0.53	2.39	0.28
	Count	17	101	17

*Considering screening data for groundwater-fed open ditches collected in summer 2003 only, the nitrate-nitrogen concentration for these open ditches is higher than the corresponding one for the River Eider.

Note: The small watercourses include ditches located within the River Eider Valley. Watercourse data were collected in May and June between 1999 and 2003. River Eider data were collected in May and June between 1999 and 2002. Ammonia-nitrogen, nitrite-nitrogen and ortho-phosphate-phosphorus concentrations have also been measured, but their mean concentrations were <0.1 mg/l.

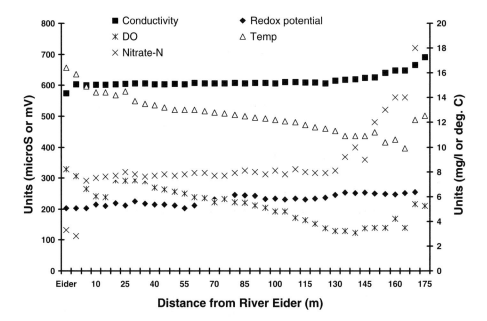

Fig. 31.5 Comparison of standard water quality variables (easy and cost-efficient to measure) and nitrate-nitrogen (difficult and cost-intensive to measure) for Ditch 24 on 22 July 2003. Single data values have been presented.

A broad suite of parameters was analysed in 24 ditches. Mean BOD, ammonia-nitrate, nitrite-nitrogen, nitrate-nitrogen and ortho-phosphate-phosphorus concentrations were 6.0, 0.2, 0.0, 1.6 and 0.0 mg/l, respectively. Table 31.2 shows the summary statistics for the BOD and nutrient concentrations of different stretches of the selected Ditches 22, 23 and 24, which are statistically significantly different from each other (see below). Elevated BOD and nitrate concentrations for upstream groundwater-fed ditch cross-sections were apparent (Table 31.2) regardless of the observation that all ditches belonged to different classes (Fig. 31.2). A spatial pattern is apparent where the nitrate-nitrogen concentration declined by about 60% within the ditch. Ditches that were not groundwater-fed (Fig. 31.2) always had nitrate-nitrogen values <2.5 mg/l (data not shown; method described above).

Figures 31.6–31.8 show comparisons of the conductivity, redox potential and DO concentrations for selected ditches with distances from their confluence with the River Eider. Obvious statistical data variations within and between most ditches were noted. Considering conductivity, Ditch 22 was associated with relatively low values (Fig. 31.6).

Figures 31.7 and 31.8 show relatively low redox potentials and DO concentrations for Ditches 22 and 23. The mechanisms responsible for the slight decrease of the redox potential for Ditch 23 have not been investigated. However, it has to be noted that Ditch 23 had the highest discharge despite of a small channel profile (Fig. 31.2).

An ANOVA for the Ditches 14, 22, 23 and 24 with respect to the variables conductivity (Fig. 31.6), redox potential (Fig. 31.7), DO (Fig. 31.8) and pH was performed. The findings confirmed that the ditches were statistically significantly different in terms of their standard water quality characteristics ($P < 0.01$). This explains the high standard deviations in Table 31.2. However, the analysis was only performed for the lower 90 m of each ditch (19 cross-sections in total), because the lower stretches always provided sufficient sample water.

Figure 31.5 shows a comparison between standard water quality variables and nitrate-nitrogen for Ditch 24. According to the German river classification scheme (Landesarbeitsgemeinschaft Wasser, 1998), the upper stretches of the ditches are classified as 'highly contaminated' (between 10 and 20 mg/l nitrate-nitrogen) due to the high nitrate contamination. The correlation coefficients between nitrate-nitrogen, and the redox potential and temperature were 0.62 and −0.55, respectively, and are statistically significant ($P < 0.05$).

31.4.3 Ditch vegetation

The overall ditch vegetation was dominated by tall and productive plant species such as *Phalaris arundinacea* L. and *Carex acutiformis* EHRH. These species indicate nutrient-rich conditions. The species composition of the ditches reflected

Table 31.2 Summary statistics for the unfiltered BOD[1] and filtered nutrient concentrations of different stretches of the Ditches 22, 23 and 24 (combined data for 9 July 2003)

Statistics	Mean	SD[2]	Mean	SD[2]	Mean	SD[2]	Mean	SD[2]	Mean	SD[2]	Mean	SD[2]
Stretch	0 m (near Eider)		30 m		60 m		90 m		120 m		150 m (near spring)	
BOD[1] (mg/l)	6.0	2.74	1.8	0.40	5.2	2.39	2.0	1.75	14.6	1.76	14.8	ned[3]
Ammonia-nitrogen (mg/l)	0.13	0.081	0.07	0.037	0.09	0.039	0.09	0.023	0.13	0.030	0.13	0.050
Nitrate-nitrogen (mg/l)	2.8	0.33	5.0	2.99	4.2	1.92	4.8	1.86	5.3	1.32	13.1	2.01

[1] Seven-days @ 20 °C ATU (nitrification inhibitor) biochemical oxygen demand (BOD); [2] standard deviation; [3] not enough data (only one value for Ditch 24).

Note: The water level for all corresponding sections was ≥ 3 cm. Nitrite-nitrogen and ortho-phosphate-phosphorus concentrations have also been measured, but their mean concentrations were below 0.1 mg/l.

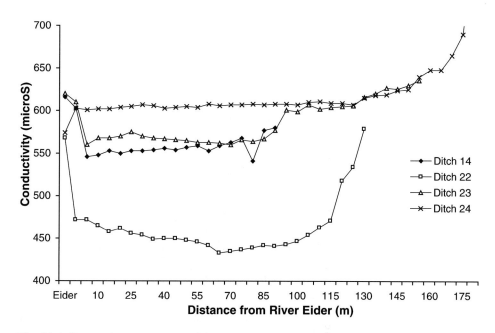

Fig. 31.6 Comparison of the conductivity for the selected Ditches 14, 22, 23 and 24. The low conductivity of Ditch 22 has been influenced by dilution via a wetland (i.e. cut-off former river stretch) in the North that is associated with low conductivity values. The conductivity axis starts at 400 μS to highlight data variation.

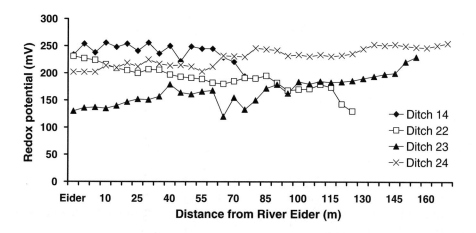

Fig. 31.7 Comparison of the redox potential for the selected Ditches 14, 22, 23 and 24. Redox potentials were only recorded for water levels ≥4 cm.

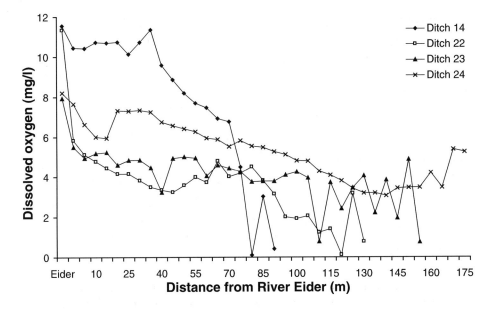

Fig. 31.8 Comparison of the dissolved oxygen (DO) for the selected Ditches 14, 22, 23 and 24.

the land use intensity of the adjacent fields. All four selected ditches were dominated by *P. arundinacea*. However, the growth of *C. acutiformis* EHRH. was also
dominant for the most extensively used Ditch 14 (Fig. 31.9).

Furthermore, the natural flow obstruction cover was at least 70% upstream
of cross-section 60 m for all selected ditches. Elevated flow obstruction cover
percentages after ditch clearance (manual vegetation removal) were merely due
to the development of meander within the silt bed upstream of cross-section 100 m
(data not shown).

31.5 Discussion

31.5.1 Disappearance of ditches due to vegetation growth

The River Eider Valley is characterised by a mosaic of different fields, all of
them have been drained in the past. In recent years, the authorities have promoted
ceasing of both mowing of ditch vegetation and dredging of ditches to encourage
the rise of the water table.

Due to the deliberate absence of an active overall ditch maintenance plan,
≥85% of all watercourses in the River Eider Valley were small in terms of their
width (<150 cm) and depth (<100 cm). Greater than 75% of all watercourses

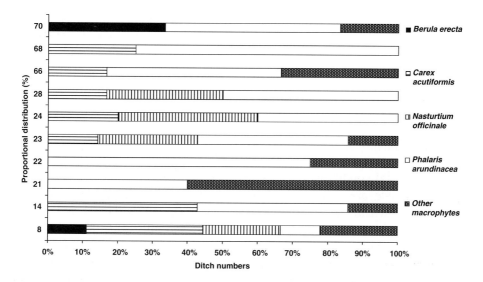

Fig. 31.9 Comparison of the dominant macrophytes within or near (1 m of each side of the embankments) 10 randomly selected ditches (160 watercourses including 39 ditches in total) of the study area.

were heavily vegetated, and had therefore flow obstruction covers of $\geq 50\%$ in May 2003 (Fig. 31.2). Nevertheless, approximately 25% of all watercourses were classified as individual ditches with a major drainage purpose (average discharge >0.5 l/s). However, ditches with a low discharge still have a minor drainage function and influence the nearby diversity and composition of vegetation.

All watercourses were small (average depth below 0.8 m) and densely vegetated, because they have not been managed (e.g., ceasing of mowing and dredging) for up to 10 years. The agricultural land was only used extensively for cattle and horse grazing. It follows that predominantly species from the Phragmitetalia australis group (macrophytes like reeds) and rushes grow along the sides of the ditches, and a very dense overgrowth of aquatic plants in the centre of the crosssection of the ditches can be noted (Fig. 31.9). The corresponding estimated values were therefore generally $\ll 5$ m$^{1/3}$ $\times$ s (International Institute for Land Reclamation and Improvement, 1964).

The water quality analysis shows a great variation of nutrient concentrations within the different ditches (SD shown in Table 31.1). However, high carbon and nutrient losses (particularly nitrate-nitrogen) to the drainage water is apparent (Table 31.1). A discussion on the effective monitoring of nitrate-nitrogen change is beyond the scope of this chapter, but can be found elsewhere (Bockholt et al., 1992; Burt, 2003).

As the ditches continue to overgrow, it is likely that there will be virtually no more ditches with a significant drainage function in the River Eider Valley in about a decade. The increase of the water table will create a shallow freshwater lake near the river, and a groundwater-fed buffer zone between the hillslope and the river. Both developments will reduce further peat degradation at higher locations.

31.5.2 Water quality variations within the study area

The hydrological conditions in the study area between the villages Techelsdorf and Schmalstede are characterised by groundwater flow from the upwelling spring area down to the River Eider (Jensen et al., 2001; Trepel and Kluge, 2002) through predominantly vegetated ditches and silted-up tile drains. The presence of the ditches reduces nutrient retention and biochemical transformation (e.g., denitrification as described by Mohamed et al. (2003)) within the peatland. The rewetting of the ditch embankments and peat soils increased the potential of sedge reeds to accumulate nutrients in the peat. The water quality analysis showed great variation of nutrients within the different ditches (Figs. 31.5 and 31.6).

A comparison between the overall water quality of the River Eider and the water quality of all small watercourses indicated that the total discharge from the small watercourses usually does not pollute the River Eider in terms of nutrient input (Table 31.1). This is partially the case due to the good water quality (except for high nitrate contamination) of the groundwater and the presence of heavy aquatic vegetation that takes up nutrients. As the ditches continue to overgrow and the water table continues to increase, it is likely that nitrate-nitrogen will be reduced further. The inflow from the ditches located in the River Eider Valley into the River Eider contributes to approximately 10% of the overall River Eider discharge (Trepel and Kluge, 2004).

Nitrate-nitrogen concentrations above 1 mg/l (see also Table 31.1) are usually an indication for man-made pollution; e.g., nitrate-nitrogen from fertilized agricultural fields located on nearby hills; peatland use for cattle grazing purposes requires drainage and fertilisation. Nitrate concentrations were higher for grazed (Ditches 23 and 24: 3.5 to 6.5 mg/l) than ungrazed ditches (Ditches 14 and 22: 0.1 to 0.7 mg/l). Peatlands have a high denitrification potential (Davidson et al., 2002), and are therefore suitable for natural pollution load reduction.

Elevated conductivity concentrations for ditches in the River Eider Valley are usually associated with diffuse environmental pollution such as urban and agricultural runoff. The conductivity within all ditches was fairly constant (except for elevated levels near the springs) but on different concentration levels for each ditch. For example, the moderately vegetated and wide Ditch 22 was associated with relatively low values (450 µS) in contrast to Ditches 14, 23 and

24 (550–600 μS). This observation can be explained by the presence of a high productive wetland (i.e. cut-off former stretch of the River Eider) in the North of Ditch 22. However, further studies on this wetland were beyond the scope of this study.

The redox potentials of Ditches 14 and 23 increased from the springs towards the river. The opposite trend was apparent for Ditches 22 and 24 (Fig. 31.7). A transformation of SO_4^{2-} to H_2S ('rotten egg' smell was not detected) may take place well below 180 mV.

Most representative ditches show typical characteristics of groundwater-fed ditches with dominant macrophyte communities (see below). Groundwater-fed ditches are associated with a constant temperature, relatively high nitrate-nitrogen concentration, relatively low oxygen concentration, low siltation and a relatively moderate flow obstruction cover. Usually, these groundwater-fed ditches are deep, have a constant water level and are associated with moderate temperature fluctuations. Furthermore, these ditches are usually mesotroph to eutroph in lowlands such as the River Eider Valley.

31.5.3 Vegetation characterisation

Table 31.3 gives an overview of the vegetation characteristics of the selected Ditches 14, 22, 23 and 24. The ditch vegetation is dominated by tall and productive plant species such as *P. arundinacea* L., *C. acutiformis* EHRH., *Glyceria maxima* (HARTM.) HOLMB. or *Urtica dioica* L. s. str.. These species indicate nutrient rich conditions (Wisskirchen and Häupler, 1998). Species typical for less nutrient rich conditions (*Carex paniculata* L. or *Carex cespitosa* L.) occur only rarely at the top and middle sections of ditches.

The ditches referred to in Fig. 31.9 have been selected at random to give an unbiased overview of macrophyte communities associated with the ditches of the River Eider Valley. Figure 31.9 indicates the cover and dominance of the Phragmitetea australis group (macrophytes) close to the drainage ditches. This is particularly the case for Ditches 14, 23, 24 and 66, for example. *Phalaris arundinacea* (Fig. 31.9) stands are typical for drained peat soil. These stands predominantly occur near the riverbed and the lower end of the ditches (Ditches 21, 29, 33 and 69) where flooding results in higher water levels.

The growth of submerged *S. emersum* L. in the River Eider leads to localised flooding during storm events. However, a cluster analysis revealed no significant link between different ditch groups and stretches influenced by different hydrological conditions and aquatic plant species. This might be explained with the observation that most plant species were cosmopolitan. However, a detailed discussion of the influence of flooding and other hydrological changes on the vegetation community dynamics is beyond the scope of this chapter.

Table 31.3 Vegetation characteristics of selected representative ditches (Ditches 14, 22, 23 and 24); Nomenclature after Wisskirchen and Häupler (1998). Classes were indicated as rare = 1 (<1% cover), present = 2 (1 – 30% cover) and dominant = 3 (>30% cover)

Ditch number/ distance to River Eider (m) Species	Group	14 0	14 50	14 100	22 0	22 50	22 100	23 0	23 50	23 100	24 0	24 50	24 100
Poa trivialis L. s. l.	MA[1]					2	3					1	
Ranunculus repens L.	MA[1]	1	1										
Caltha palustris L.	MO[2]							1	1				
Carex cespitosa L.	MO[2]	2	2					1					
Epilobium hirsutum L.	MO[2]			2									
Filipendula ulmaria (L.) MAXIM.	MO[2]		2	1		2	2		2	1			
Juncus effusus L.	MO[2]		1				2						
Lysimachia nummularia L.	MO[2]								1			1	
Myosotis scorpioides agg.	MO[2]				2			1	1		2		
Scirpus sylvaticus L.	MO[2]						2						
Berula erecta (HUDS.) COVILLE	PH[3]		1		1								
Calamagrostis canescens agg.	PH[3]		2										
Carex acutiformis EHRH.	PH[3]	2	3	2					2			2	
Carex paniculata L.	PH[3]			2		3			2				
Equisetum fluviatile L.	PH[3]					1	2						
Galium palustre L. s. l.	PH[3]		1						1			1	
Glyceria maxima (HARTM.) HOLMB.	PH[3]		3		1				1				
Nasturtium officinale R. Br.	PH[3]			1	1			2	3		2		3
Persicaria hydropiper (L.) DELARBRE	PH[3]	1			2			2	1		2	1	
Phalaris arundinacea L.	PH[3]	3	3	2	3	2	2	3	3	3	3	3	1
Rorippa palustris (L.) BESSER	PH[3]				1						1		
Stellaria alsine GRIMM	PH[3]	1			1						1		
Valeriana officinalis L. s. str.	PH[3]		1						1				
Galeopsis tetrahit L.	AR[4]				1	1					1		
Galium aparine L.	AR[4]		1	1	1	1	1					2	
Urtica dioica L. s. str.	AR[4]	1		2		2	2					3	
Alopecurus geniculatus L.	LP[5]				1			1					
Glyceria fluitans (L.) R. BR.	LP[5]		1		1								
Lemna minor L.	LE[6]				2			2					

[1] Molinio-Arrhenatheretea.
[2] Molinietalia.
[3] Phragmitetea australis.
[4] Artemisietea vulgaris.
[5] Lolio-Potentillion.
[6] Lemnetea.

31.5.4 Hydraulic changes due to summer flooding

Groundwater-fed ditches were less overgrown than ditches supplied by surface runoff only (Fig. 31.2). In fact, most ditches that were not fed by groundwater and first dredged in the nineteenth century have disappeared (Trepel and Kluge, 2002). A two-way flow between ditch water and groundwater was not significant for the upper ditch stretches, considering discharge values of <2 l/s and that multiple springs were clearly visible in each selected ditch. However, an interaction of ditch, river and groundwater was likely during flooding.

Studies in the River Eider Valley have shown that the water quality within groundwater-fed ditches is relatively constant over time for the season sampled, with the exception of the stretch near the River Eider where mixing between ditch and Eider water takes place. Water quality variables such as redox potential and conductivity did not change significantly during flooding.

Mixing at the lower ditch stretch was due to the fluctuations of the River Eider water level (approximately 80 cm per annum). In general, the reasons for high water levels during summer and winter are vegetation growth (predominantly *S. emersum* L.) and precipitation (increased discharge from upstream urban regions), respectively.

For the headwater zone of the upper River Eider, the slope usually drains directly into the river channel. A small floodplain separating slope and river channel is located further downstream in the River Eider Valley. The hydrological function of the floodplain in the valley is predominantly to provide a conduit for slope drainage. Moreover, land drainage via ditches and tile drains connects the slope to the channel. It follows that the existence of the floodplain is of little hydraulic importance, because slope drainage short-circuits the floodplain. However, heavily vegetated and silted-up ditches contribute, together with the vegetated River Eider, to the flooding of the adjacent floodplains.

Ceasing macrophyte mowing resulted in localised flooding of the embankments of the River Eider. For example, Fig. 31.4a shows a comparison of the water level near the outlet arrangement of selected ditches before and during summer flooding in the absence of significant precipitation. This finding confirms observations made earlier by Jensen et al. (2001), and Trepel and Kluge (2002).

31.6 Conclusions

The findings of the watercourse classification exercise indicated the dominance of small heavily-vegetated ditches. A minority of predominantly large ditches was supplied with groundwater at their upper stretches. The ceasing of aquatic plant mowing resulted in heavy vegetation growth within both the selected ditches and the River Eider leading to a rise in the water level of approximately 0.8 m.

It is likely that ceasing ditch and river maintenance will lead to the disappearance of the ditches and subsequently to a rise of the water table in the neighbouring agricultural fields, which is intended by the federal state government. This strategy should result in the improvement of the water quality of the River Eider. The likely formation of a shallow freshwater lake in the River Eider Valley will subsequently lead to the flooding of the riparian wetland, and to a decline of peat degradation.

The natural horizontal flow obstruction area cover was at least 70% upstream of cross-section 60 m for the four selected representative ditches. Elevated flow obstruction cover values after ditch clearance were merely due to the development of meanders within the silt bed upstream of cross-section 100 m. The total-roughness values of both the vegetated and excavated ditch stretches were relatively low in comparison to values for well-maintained open channels reported in the hydraulic literature. However, it would be difficult for an inexperienced scientist or engineer to correctly identify values by eye for the different ditch classes. Moreover, the total-roughness is also a function of season influencing vegetation growth.

Significant hydrochemical differences within the selected ditches were noted. For example, the DO and water temperature within the ditch waters increased along the ditches towards the River Eider. Elevated BOD and nitrate-nitrogen concentrations for upstream ditch cross-sections were also apparent. Peat degradation and groundwater infiltration were likely sources of diffuse pollution.

Statistically significant water quality data variations between most ditches were also noted. For example, elevated conductivity concentrations indicated usually diffuse environmental pollution. The conductivity within all ditches was fairly constant (except for elevated levels near the springs), but on different concentration levels. Ditches that were not dominated by groundwater infiltration had always nitrate-nitrogen values <2.5 mg/l. Nitrate concentrations were higher for grazed than ungrazed ditches.

A comparison between the water qualities of the River Eider and a representative selection of small watercourses indicated that the overall discharge from the small watercourses usually does not pollute the River Eider in terms of their average nutrient concentrations during May and June. This finding shows that ceasing ditch maintenance has no negative impact on the overall water quality of the River Eider.

The River Eider flooded large parts of the riparian wetland, and subsequently discharged into the ditches during late summer. This was due to the growth of *S. emersum* L. in the River Eider leading to an increase of the water level. However, the growth of vegetation within the ditches also resulted in elevated water levels. It follows that the floodplains were flooded. The creation of a temporary shallow freshwater lake is anticipated, if ceasing of watercourse maintenance continues for another decade.

In particular, the Phragmitetea australis group (tall and productive macrophytes) dominated the ditch ecosystem. All four selected ditches were heavily vegetated by *P. arundinacea* L. Moreover, the growth of *C. acutiformis* EHRH. was also dominant for the most extensively used ditch. Nevertheless, the wetland vegetation communities near the ditches could not be used meaningfully to further classify the identified groups of ditches within a water engineering and drainage context. No statistical data link between the cosmopolitan vegetation and the hydrology of the selected ditches was apparent even ten years after ceasing ditch maintenance. Further work on indicator plant species is going on.

References

A. Aamodt and E. Plaza, "Case-based reasoning: foundational issues, methodological variations and system approaches", AI Communications, 7 (1), 39–59 (1994).

C.L. Abbott and L. Comino-Mateos, "In situ performance monitoring of an infiltration drainage system and field testing of current design procedures", J. Chart. Inst. Wat. Environ. Manage., 15 (3), 198–202 (2001).

C. Ahn, W.J. Mitsch and W.E. Wolfe, "Effects of recycled FGD liner material on water quality and macrophytes of constructed wetlands: a mesocosm experiment", Wat. Res., 35 (3), 633–642 (2001).

P.A. Aguilera, A.G. Frenich, J.A. Torres, H. Castro, J.L.M. Vidal and M. Canton, "Application of the Kohonen Neural Network in coastal water management: methodological development for the assessment and prediction of water quality", Wat. Res., 35 (17), 4053–4062 (2001).

S.E. Allen, Chemical Analysis of Ecological Materials (Blackwell, Oxford, UK, 1974).

B.J. Alloway, Heavy Metals in Soils, 2nd ed. (Chapman and Hall, Suffolk, UK, 1995).

R.Y.G. Andoh and C. Declerck, "A cost effective approach to stormwater management? Source control and distributed storage", Wat. Sci. Tech., 36 (8–9), 307–311 (1997).

Y. Ann, K.R. Reddy and J.J. Delfino, "Influence of chemical amendments on phosphorus immobilization in soils from a constructed wetland", Ecol. Eng., 14 (1–2), 157–167 (1999a).

Y. Ann, K.R. Reddy and J.J. Delfino, "Influence of redox potential on phosphorus solubility in chemically amended wetland organic soils", Ecol. Eng., 14 (1–2), 169–180 (1999b).

D. Arditi and O.B. Tokdemir, "Comparison of case-based reasoning and artificial neural networks", J. Comp. Civ. Eng., 13 (3), 162–168 (1999).

B. Arheimer and M. Brandt, "Modelling nitrogen transport and retention in the catchments of southern Sweden", Ambio., 27 (6), 471–480 (1998).

S.O. Astebøla, T. Hvitved-Jacobsenb and O. Simonsen, "Sustainable stormwater management at Fornebu - from an airport to an industrial and residential area of the city of Oslo, Norway", Sci. Tot. Env., 334–335, 239–249 (2004).

ATV-DVWK-Arbeitsgruppe, Planung, Bau und Betrieb von Anlagen zur Versickerung von Niederschlagswasser (Design, Construction and Operation of Rainwater Drainage Systems), Regelwerk A-138, ES-41, ISBN 3-935669-83-6, (ATV-DTWK (German Association for Water, Wastewater and Waste), Gesellschaft zur Förderung der Abwassertechnik e.V., Hennef, Germany, 2002) (in German).

P.A.M. Bachand and A.J. Horne, "Denitrification in constructed free-water surface wetlands: I. Very high nitrate removal rates in a macrocosm study", Ecol. Eng., 14 (1–2), 9–15 (1999).

M. Backstrom, S. Karlsson, L. Backman, L. Folkeson and B. Lind, "Mobilisation of heavy metals by deicing salts in a roadside environment", Wat. Res., 38 (3), 720–732 (2004).

A.S. Ball, M. William, D. Vincent and J. Robinson, "Algae growth control by barley straw extract", Biores. Technol., 77 (2), 177–181 (2001).

L. Belanche, J.J. Valdes, J. Comas, I.R. Roda and M. Poch, "Prediction of the bulking phenomenon in wastewater treatment plants", Artif. Intell. Eng., 14 (4), 307–317 (2000).

R. Bettess, "Infiltration drainage - manual of good practice (R156)", Construction Industry Research and Information Association (CIRIA) Report 156, ISBN: 0-86017-457-3 (CIRIA, London, UK, 1996).

R.R. Boar, C.E. Crook and B. Moss B, "Regression of *Phragmites australis* reedswamps and recent changes of water chemistry in the Norfolk Broadland, England", Aquat. Bot., 35 (1), 41–55 (1989).

R. Bockholt, G. Koch, W. Ebert and E. Fedderwitz, "Nutrient contents of drainwater and drainage ditchwater on agriculturelly used areas in the drinking water territory of the River Kösterbeck", Zeitschrift für Kulturtechnik und Landentwicklung, 33, 178–185 (1992) (in German).

P.I. Boon and A. Mitchell, "Methanogenesis in the sediments of an Australian freshwater wetland: comparison with aerobic decay and factors controlling methanogenesis", FEMS Microbiol. Ecol., 18 (3), 174–190 (1995).

J. Braun-Blanquet, Plant Sociology. In: G.D. Fuller and H.C. Conrad (eds.), The Study of Plant Communities (McGraw-Hill, New York, USA, 1932).

H. Brix, "Do macrophytes play a role in constructed treatment wetlands"? Wat. Sci. Tech., 35 (5), 11–17 (1997).

H. Brix, "Functions of macrophytes in constructed wetlands", Wat. Sci. Tech., 29 (4), 71–78 (1999).

H. Brix, C.A. Arias and M. del Bubba, "Media selection for sustainable phosphorus reduction in subsurface flow constructed wetlands", Wat. Sci. Tech., 44 (11–12), 47–54 (2001).

Building Research Establishment, Soakaway Design, Building Research Establishment (BRE) Digest 365, which replaced BRE Digest 151 (BRE Bookshop, Watford, UK, 1991).

T. Bulc and A.S. Slak, "Performance of constructed wetland for highway runoff treatment", Wat. Sci. Tech., 48 (2), 315–322 (2003).

T.P. Burt, "Monitoring change in hydrological systems", Sci. Tot. Env., 310 (1–3), 9–16 (2003).

D. Butler and J.W. Davies, Urban Drainage (E & FN Spon, London, UK, 2000).

D. Butler and J. Parkinson, "Towards sustainable urban drainage", Wat. Sci. Tech., 35 (9), 53–63 (1997).

D. Butler, Y. Xiao, S.H.P.G. Karunaratne and S. Thedchananamoorthy, "The gully pot as a physical, chemical and biological reactor", Wat. Sci. Tech., 31 (7), 219–228 (1995).

C.S. Campbell and M.H. Ogden, Constructed Wetlands in the Sustainable Landscape (John Wiley & Sons, New York, USA, 1999).

L.J. Cao and F.E.H. Tay, "Support Vector Machine with adaptive parameters in financial time series forecasting", IEEE Transactions on Neural Networks, 14 (6), 1506–1518 (2003).

T. Caquet, L. Lagadic, O. Jonot, W. Baturo, M. Kilanda, P. Simon, S. LeBras, M. Echaubard and F. Ramade, "Outdoor experimental ponds (mesocosms) designed for long-term ecotoxicological studies in the aquatic environment", Ecotoxicol. Environ. Saf., 34 (2), 125–133 (1996).

R. Carignan and J. Kaill, "Phosphorus sources for aquatic weeds: water or sediments"? Sci., 207 (4434), 987–989 (1980).

G.A. Carpenter and N. Markuzon, "ARTMAP-IC and medical diagnosis: instance counting and inconsistent cases", Neural Networks, 11 (2), 323–336 (1998).

R. Cereghino, J.L. Giraudel and A. Compin, "Spatial analysis of stream invertebrates distribution in the Adour-Garonne Drainage Basin (France) using Kohonen Self-organizing Maps", Ecol. Model., 146 (1–3), 167–180 (2001).

C.-C. Chang and C.-J. Lin (2005), LIBSVM: a Library for Support Vector Machines. Software available at http://www.csie.ntu.edu.tw/~cjlin/libsvm.

M.H. Chaudhry, Principles of Flow of Water. In: L.W. Mays (ed.), Resources Handbook (McGraw-Hill, New York, USA, 1996).

S.P. Cheng, W. Grosse, F. Karrenbrock and M. Thoennessen, "Efficiency of constructed wetlands in decontamination of water polluted by heavy metals", Ecol. Eng., 18 (3), 317–325 (2002).

CIRIA, Sustainable Urban Drainage Systems: Design Manual for Scotland and Northern Ireland, Construction Industry Research and Information Association (CIRIA) Report C521 (Cromwell Press, London, UK, 2000).

City of Edinburgh Council, Flood Assessment Report – Edinburgh Flood Assessment Study (Edinburgh City Development Department, Edinburgh, Scotland, UK, 1999).

L.S. Clesceri, A.E. Greenberg and A.D. Eaton, Standard Methods for the Examination of Water and Wastewater, 20th ed. (American Public Health Association, American Water Works Association and Water Environment Federation, Washington DC, USA, 1998).

Convention on Wetlands of International Importance Especially as Waterfowl Habitat (Ramsar, Iran, 1971).

P.F. Cooper, G.D. Job, M.B. Green and R.B.E. Shutes, Reed Beds and Constructed Wetlands for Wastewater Treatment (WRc, Swindon, UK, 1996).

Council of European Communities, "Directive of 23 October 2000 establishing a framework for community action in the field of water policy (2000/60/EC)", Official Journal, L327, 0001–0073 (2000).

C.R. Curds, Protozoa in the Water Industry (Cambridge University Press, Cambridge, UK, 1992)

B.J. D'Arcy and A. Frost, "The role of best management practices in alleviating water quality problems associated with diffuse pollution", Sci. Tot. Env., 265 (1), 359–367 (2001).

T.E. Davidson, M. Trepel and J. Schrautzer, "Denitrification in drained and rewetted minerotrophic peat soils in Northern Germany (Pohnsdorfer Stauung)", J. Plant Nutr. Soil Sci., 165, 199–204 (2002).

T.N. Debo and A.J. Reese, Municipal Stormwater Management, 2nd ed. (Lewis Publishers, New York, USA, 2003).

DEFRA, Development and Flood Risk – Joint Report to the Department for Environment, Food and Rural Affairs (DEFRA) and the Office of the Deputy Prime Minister (Environment Agency, London, UK, 2004).

Deutsches Institut für Normung, DIN EN ISO 13395: Water Quality – Determination of Nitrite Nitrogen and Nitrate Nitrogen and the Sum of Both by Flow Analysis (CFA and FIA) and Spectrometric Detection (Deutsches Institut für Normung e.V., Beuth Verlag, Berlin, Germany, 1996a).

Deutsches Institut für Normung, DIN EN 1189: Wasserbeschaffenheit - Bestimmung von Phosphor - Photometrisches Verfahren mittels Ammoniummolybdat (Deutsches Institut für Normung e.V., Beuth Verlag, Berlin, Germany, 1996b) (in German).

Deutsches Institut für Normung, DIN EN ISO 11732: Wasserbeschaffenheit - Bestimmung von Ammoniumstickstoff - Verfahren mittels Fließanalyse (CFA und FIA) und spektrometrischer Detektion (ISO/DIS 11732:2003) (Deutsches Institut für Normung e.V., Beuth Verlag, Berlin, Germany, 2003) (in German).

B. Dong, C. Cao and S.E. Lee, "Applying support vector machines to predict building energy consumption in tropical regions", Energ. Build., 37 (5), 545–553 (2005).

N. Dong, W.C. Lu, N.Y. Chen, Y.C. Zhu and K.X. Chen, "Using support vector classification for SAR of fentanyl derivatives", Acta Pharmacol. Sin., 26 (1), 107–112 (2005).

V.A. Donkor and D.-P. Häder, "Effects of ultraviolet radiation on photosynthetic pigments in some filamentous cyanobacteria", Aquat. Microb. Ecol., 11 (2), 143–149 (1996).

A. Dubey, M.J. Realff, J.H. Lee and A.S. Bommarius, "Support vector machines for learning to identify the critical position of a protein", J. Theor. Biol., 234 (3), 351–361 (2005).

D. Dubois and H. Prade, "An introduction to fuzzy systems", Clin. Chim. Acta, 270 (1), 3–29 (1998).

W. Duch and K. Grudzinski, Weighting and Selection of Features, Proceedings of Intelligent Information Systems VIII Workshop, Ustron, Poland, June 1999, pp. 32–36.

T. Dunne and L.B. Leopold, Water in Environmental Planning (W.H. Freeman and Company, New York, USA, 1978).

J.B. Ellis, B.J. D'Arcy and P.R. Chatfield, "Sustainable urban drainage systems and catchment planning", J. Chart. Inst. Wat. Environ. Manage., 16 (4), 286–291 (2002).

J.B. Ellis, J-C. Deutsch, J-M. Mouchel, L. Scholes and M.D. Revitt, "Multicriteria decision approaches to support sustainable drainage options for the treatment of highway and urban runoff", Sci. Tot. Env., 334–335, 251–260 (2004).

J.B. Ellis, "Integrated approaches for achieving sustainable development of urban storm drainage", Wat. Sci. Tech., 32 (1), 1–6 (1995).

J.B. Ellis, M.D. Revitt, R.B.E. Shutes and J.M. Langley, "The performance of vegetated bio-filters for highway runoff control", Sci. Tot. Env., 146–147 (5), 543–550 (1994).

J.B. Ellis, R.B.E. Shutes and M.D. Revitt, Constructed wetlands and links with sustainable drainage systems. Technical Report P2-159/TR1 (Environment Agency, Bristol, UK, 2003).

EPA, StormWater Technology Fact Sheet - Wet Detention Pond, 832-F-99-048, (United States Environmental Protection Agency (EPA), Office of Water, Washington DC, USA, 1999).

P.G. Eriksson and S.E.B. Weisner, "An experimental study on effects of submersed macrophytes on nitrification and denitrification in ammonium-rich aquatic systems", Limnol. Oceanogr., 44, (1999).

J.R. Etherington, Wetland Ecology (Edward Arnold, London, UK, 1983).

European Environment Agency, Environment in the European Union at the Turn of the Century, Environmental Assessment Report 2 (European Environment Agency, Copenhagen, Denmark, 1999).

N.C. Everall and D.R. Lees, "The identification and significance of chemicals released from decomposing barley straw during reservoir algal control", Wat. Res., 31 (3), 614–620 (1997).

S.P. Faulkner and C.J. Richardson, Physical and chemical characteristics of freshwater wetland soils, In: D.A. Hammer (ed.), Constructed Wetlands for Wastewater Treatment (Lewis Publishers, Chelsea, MI, USA, 1989) pp. 41–72.

M.S. Fennessy, C.C. Brueske, W.J. Mitsch, "Sediment deposition patterns in restored freshwater wetlands using sediment traps", Ecol. Eng., 3 (4), 409–428 (1994).

R. Field and D. Sullivan, Wet-Weather Flow in the Urban Watershed: Technology and Management (Lewis Publishers, New York, USA, 2003).

J. Fowler and L. Cohen, Practical Statistics for Field Biology (John Wiley and Sons, West Sussex, UK, 1998).

H.-G. Frede and S. Dabbert, Handbuch zum Gewässerschutz in der Landwirtschaft (Ecomed, Landsberg, Germany, 1998) (in German).

L.H. Fredrickson and F.A. Reid, Impacts of hydrologic alteration on management of freshwater wetlands", In: J.M. Sweeney (ed.), Management of Dynamic Ecosystems (North Central Section, Wildlife Society, West Lafayette, Indiana, USA, 1990) pp. 71–90.

R. Gächter and J.S. Meyer, "The role of micro-organisms in mobilization and fixation of phosphorus in sediments", Hydrobiol., 253 (6), 103–121 (1993).

R. Gächter, J.S. Meyer and A. Mares, "Contribution of bacteria to release and fixation of phosphorus in lake sediments", Limnol. Oceanogr. 33 (6), part 2, 1542–1558 (1988).

M.R. Galuzzi and J.M. Pflaum, "Integrating drainage, water quality, wetlands, and habitat in a planned community development", J. Urban Plng. and Devel., 122 (3), 101–108 (1996).

R.P. Gambrell and W. H. Patrick Jr., Chemical and microbiological properties of anaerobic soils and sediments, In: D.D. Hook and R.M.M. Crawford (eds.), Plant Life in Anaerobic Environments (Ann Arbor Science, Ann Arbor, 1978) pp. 375–423.

H.L. Garcia and I.M. Gonzalez, "Self-organizing map and clustering for wastewater treatment monitoring", Eng. Appl. Artif. Intel., 17 (3), 215–225 (2004).

K.V. Gernaey, M.C.M. van Loosdrecht, M. Henze, M. Lind and S.B. Jørgensen, "Activated sludge wastewater treatment plant modeling and simulation: state of the art", Environ. Modell. Softw., 19 (9), 763–783 (2004).

L. Gervin and H. Brix, "Reduction of nutrients from combined sewer overflows and lake water in a vertical-flow constructed wetland system", Wat. Sci. Tech., 44 (11–12), 171–176 (2001).

M.O. Gessner, "Breakdown and nutrient dynamics of submerged *Phragmites* shoots in the littoral zone of a temperate hardwater lake", Aquat. Bot., 66 (1), 9–20 (2000).

M. Gevrey, F. Rimet, Y.S. Park, J.L. Giraudel, L. Ector and S. Lek, "Water quality assessment using diatom assemblages and advanced modelling techniques", Freshwat. Biol., 49 (2), 208–220 (2004).

C. Gleeson and N. Gray N., The Coliform Index and Waterborne Disease (E & FN Spon, London, UK, 1997).

M. Green, E. Friedler and I. Safrai, "Enhancing nitrification in vertical-flow constructed wetlands utilizing a passive air pump", Wat. Res., 32 (12), 3513–3520 (1998).

S.V. Gregory, F.J. Swanson, W.A. McKee and K.W. Cummins, "An ecosystem perspective of riparian zones", Biosci., 41 (8) 540–551 (1991).

S. Grieu, A. Traore, M. Polit and J. Colprim, "Prediction of parameters characterizing the state of a pollution removal biological process", Eng. Appl. Artif. Intel., 18 (5), 559–573 (2005).

P.M. Groffman, A.J. Gold and K. Addy, "Nitrous oxide production in riparian zones and its importance to national emission inventories", Chemosph.: Glob. Change Sci., 2 (3–4), 291–299 (2000).

A. Grohmann, U. Hässelbarth, W. Schwerdtfeger, Die Trinkwasserverordnung, Einführung und Erläuterungen für Wasserversorgungsunternehmen und Überwachungsbehörden, 4th ed. (Erich Schmidt Verlag, Berlin, Germany, 2002) (in German).

Y. Guo, "Hydrologic design of urban flood control detention ponds", J. Hydrol. Eng. - ASCE, 6 (6), 472–479 (2001).

M.M. Hamed, M.G. Khalafallah and E.A. Hassanien, "Prediction of wastewater treatment plant performance using artificial neural networks", Environ. Model. Softw. 19 (10), 919–928 (2004).

D.A. Hammer, Constructed Wetlands for Wastewater Treatment – Municipal, Industrial and Agricultural (Lewis Publishers, Chelsea, Michigan, USA, 1989).

G. Hanssen, Das Amt Bordesholm im Herzogtum Holstein: eine statistische monographie auf historischer grundlage (Amt Bordesholm, Kiel, Germany, 1842) (in German).

C.J. Hawke and P.V. José, Reed bed Management for Commercial and Wildlife Interests (The Royal Society for the Protection of Birds, Sandy, UK, 1996).

K. Heimburg, Hydrology of north-central Florida cypress domes, In: K.C. Ewel and H.T. Odum (eds.), Cypress Swamps (University Presses of Florida, Gainesville, Florida, USA, 1984).

R.W. Herschy, Hydrometry – Principles and Practices, 2nd ed. (John Wiley and Sons, Chichester, UK, 1998).

G.C. Holdren and D.E. Armstrong, "Factors affecting phosphorus release from intact lake sediment cores", Env. Sci. Tech., 14 (1), 79–87 (1980).

Y.-S.T. Hong, M.R. Rosen and R. Bhamidimarri, "Analysis of a municipal wastewater treatment plant using a neural network-based pattern analysis", Wat. Res., 37 (7), 1608–1618 (2003).

R.J. Hyndman and A.B. Koehler, Another Look at Measures of Forecast Accuracy, Monash Econometrics and Business Statistics Working Papers (13/05/2005).

G. Ice, "History of innovative best management practice development and its role in addressing water quality limited waterbodies", J. Envir. Eng., 130 (6), 684–689 (2004).

International Institute for Land Reclamation and Improvement, Code of Practice for the Design of Open Watercourses and Ancillary Structures (Royal van Gorcum, Assen, The Netherlands, 1964).

IWA, Constructed Wetlands for Pollution Control, International Water Association (IWA) Specialist Group 'Use of Macrophytes in Water Pollution Control' (IWA Publishing, London, UK, 2000).

C. Jefferies, A. Aitken, N. McLean, K. MacDonald and G. McKissock, "Assessing the performance of urban BMPs in Scotland", Wat. Sci. Tech., 39 (12), 123–131 (1999).

K. Jensen, O. Granke, B. Hoppe, J. Kieckbusch, M. Trepel and U. Leiner, "Weidelandschaft Eidertal – Naturschutz durch extensive Beweidung und Wiedervernässung", Petermanns Geographische Mitteilungen, 145 (1), 38–49 (2001) (in German).

T. Joachims, Making large-scale SVM learning practical, In: B. Scholkopf, C. Burges and A. Smola (eds.), Advances in Kernel Methods – Support Vector Learning, http://svmlight.joachims.org (MIT-Press, Cambridge, USA, 1999).

W.J. Junk, P.B. Bayley and R.E. Sparks, The flood pulse concept in river-floodplain systems, In: D.P. Dodge (ed.), Proceedings of the International Large River Symposium, J. Can. Fish. Aqu. Sci., 106 (special issue), 11–127 (1989).

R.H. Kadlec, "Chemical, physical and biological cycles in treatment wetlands", Wat. Sci. Tech., 40 (2), 37–44 (1999).

R.H. Kadlec, Effects of Pollutant Speciation in Treatment Wetlands Design (Wetland Management Services, Chelsea, Michigan, USA, 2002).

R.H. Kadlec and R.L. Knight, Treatment Wetlands (CRC Press, Boca Raton, Florida, USA, 1996).

R. Kadlec, R.L. Knight, J. Vymazal, H. Brix, P.F. Cooper and R. Haberl, Constructed Wetlands for Pollution Control, International Water Association (IWA) Specialist Group 'Use of Macrophytes for Water Pollution Control,' Scientific and Technical Report Number 8 (IWA Publishing, London, UK, 2000).

L. Kamp-Nielson, "Mud-water exchange of phosphate and other ions in undisturbed sediment cores and factors affecting exchange rates", Arch. Hydrobiol., 73 (2), 218–237 (1974).

A.D. Karathanasis, C.L. Potter and M.S. Coyne, "Vegetation effects on fecal bacteria, biochemical oxygen demand and suspended solids removal in constructed wetlands treating domestic wastewater", Ecol. Eng., 20 (2), 157–169 (2003).

D.S. Kaster, C.B. Medeirosand and H.V. Rocha, "Supporting modeling and problem solving from precedent experiences: the role of workflows and case-based reasoning", Environ. Model. Softw., 20 (6), 689–704 (2005).

G. Kiely, Environmental Engineering (McGraw-Hill International, Maidenhead, UK, 1997).

B. Klove, "Characteristics of nitrogen and phosphorus loads in peat mining wastewater", Wat. Res., 35 (10), 2353–2362 (2001).

R.L. Knight, R.H. Kadlec and H.M. Ohlendorf, "The use of treatment wetland for petroleum industry effluents", Environ. Sci. Technol., 33 (7), 973–980 (1999).

R. Kohavi, "A Study of cross-validation and bootstrap for accuracy estimation and model selection", Proceedings of the 14th International Joint Conference on Artificial Intelligence, 2005, pp. 1137–1143.

T. Kohonen, Self-organizing Maps, 3rd ed. (Springer Verlag, Berlin, Germany, 2001).

K.A. Kuehn, M.O. Gessner, R.G. Wetzel and K. Suberkropp, "Standing litter decomposition of the emergent macrophyte *Erianthus giganteus*", Microb. Ecol., 38 (1), 50–57 (1999).

E. Kuehn and J.A. Moore, "Variability of treatment performance in constructed wetlands", Wat. Sci. Tech., 32 (3), 241–250 (1995).

J. Kvet and S. Husak, Primary data on biomass and production estimates in typical stands of fishpond littoral plant communities, In: D. Dykyjová and J. Kvet (eds.) Pond Littoral Ecosystems (Springer Verlag, Berlin, Germany, 1978) pp. 211–216.

Länderarbeitsgemeinschaft Wasser, Beurteilung der Wasserbeschaffenheit von Fließgewässern in der Bundesrepublik Deutschland - Chemische Gewässergüteklassifikation (Kulturbuchverlag, Berlin, Germany, 1998) (in German).

K. Lee, D. Booth and P. Alam, "A comparison of supervised and unsupervised neural networks in predicting bankruptcy of Korean firms", Expert Syst. Appl., 29 (1), 1–16 (2005a).

B.-H. Lee, M. Scholz, A. Horn and A. Furber, "Constructed wetlands: prediction of performance with case-based reasoning (Part B)", Environ. Eng. Sci., 23 (2), 203–211 (2005b).

J. Leflaive, R. Cereghino, M, Danger, G. Lacroix and L. Ten-Hage, "Assessment of self-organizing maps to analyze sole-carbon source utilization profiles", J. Microbiol. Meth., 62 (1), 89–102 (2005).

P.E. Lim, M.G. Tay, K.Y. Mak and N. Mohamed, "The effect of heavy metals on nitrogen and oxygen demand reduction in constructed wetlands", Sci. Tot. Env., 301 (1–3), 13–21 (2003).

H.-X. Liu, R.-S. Zhang, X.-H. Yao, M.-C. Liu, Z.-D. Hu and B.-T. Fan, "Prediction of electrophoretic mobility of substituted aromatic acids in different aqueous-alcoholic solvents by capillary zone electrophoresis based on support vector machine", Anal. Chim. Acta, 525 (1), 31–41 (2004).

R. Lowrance, L.S. Altier, J.D. Newbold, R.R. Schnabel, P.M. Groffman, J.M. Denver, D.L. Correll, J.W. Gilliam, J.L. Robinson, R.B. Brinsfield, K.W. Staver, W. Lucas and A.H. Todd, "Water quality functions of riparian forest buffers in Chesapeake Bay watersheds", Env. Managem., 21 (5), 687–712 (1979).

R.-S. Lu and S.-L. Lo, "Diagnosing reservoir water quality using self-organizing maps and fuzzy theory", Wat. Res., 36 (9), 2265–2274 (2004).

W.-Z. Lu and W.-J. Wang, "Potential assessment of the support vector machine method in forecasting ambient air pollutant trends", Chemosphere, 59 (5), 693–701 (2005).

V. Luederits, E. Eckert, M. Lange-Weber and A. Lange, "Nutrient removal efficiency and resource economics of vertical-flow and horizontal-flow constructed wetlands", Ecol. Eng., 18 (2), 157–171 (2001).

L.J. Lund, A.J. Horne and A.E. Williams, "Estimating denitrification in a large constructed wetland using stable nitrogen isotope ratios", Ecol. Eng., 14 (1–2), 67–76 (2000).

U. Mander, V. Kuusemets and M. Ivask, "Nutrient dynamics of riparian ecotones: a case study from the Porijõgi River catchment, Estonia", Landsc. Urb. Plan., 31 (1–3), 333–348 (1995).

H.R. Maier, N. Morgan and C.W.K. Chow, "Use of artificial neural networks for predicting optimal alum doses and treated water quality parameters", Environ. Model. Softw., 19 (5), 485–494 (2004).

K.W. Mandernack, L. Lynch, H.R. Krouse and M.D. Morgan, "Sulfur cycling in wetland peat of the New Jersey Pinelands and its effect on stream water chemistry", Geochimica et Cosmochimica Acta, 22 (5), 3949–3964 (2000).

O. Mark, S. Weesakul, C. Apirumanekul, S.B. Aroonnet and S. Djordjevic, "Potential and limitations of 1D modelling of urban flooding", J. Hydro., 299 (3–4), 284–299 (2004).

P. Martin, B. Turner, K. Waddington, J. Dell, C. Pratt, N. Campbell, J. Payne and B. Reed, Sustainable Urban Drainage Systems – Design Manual for England and Wales, CIRIA Report Number C522 (Construction Industry Research and Information Association (CIRIA), London, UK, 2000).

T. Mayer and J.R. Kramer, "Effect of lake acidification on the adsorption of phosphorus by sediments", Wat. Air Soil Pol., 31 (3–4), 949–958 (1986).

P.A. Mays and G.S. Edwards, "Comparison of heavy metal accumulation in a natural wetland and constructed wetlands receiving acid mine drainage", Ecol. Eng., 16 (4), 487–500 (2001).

S.J. McNaughton, "Ecotype function in the *Typha* community-type", Ecol. Monogr., 36 (4), 297–325 (1966).

A. McNeill and S. Olley, "The effects of motorway runoff on watercourses in south-west Scotland", J. Chart. Inst. Wat. Environ. Manage., 12 (6), 433–439 (1998).

F.A. Memon and D. Butler, "Assessment of gully pot management strategies for runoff quality control using a dynamic model", Sci. Tot. Env., 295 (1–3), 115–129, 2002.

Meteorological Office, Bradford Monthly Long-term Average Rainfall, Online data between 1908 and 2001, http://www.metoffice.gov.uk/climate/uk/stationdata/bradforddata.txt, 2002.

G. Mitchell, "Mapping hazard from urban non-point pollution: a screening model to support sustainable urban drainage planning", J. Environ. Manage., 74 (1), 1–9 (2005).

W.J. Mitsch and J.G. Gosselink, Wetlands, 3rd ed. (John Wiley & Sons, New York, USA, 2000).

M.A.A. Mohamed, H. Terao, R. Suzuki, I.S. Babiker, K. Ohta, K. Kaori and K. Kato, "Natural denitrification in the Kakamigahara groundwater basin, Gifu prefecture, central Japan", Sci. Tot. Env., 307 (1–3), 191–201 (2003).

P.A. Moore and K.R. Reddy, "Role of Eh and pH on phosphorus geochemistry in sediments of Lake Okeechobee, Florida", J. Environ. Qual., 23 (5), 955–964 (1994).

M.D. Morgan and K.W. Mandernack "Biogeochemistry of sulfur in wetland peat following 3.5 Y of artificial acidification (HUMEX)", Environ. Intern., 22 (5), 605–610 (1996).

B. Moss, Ecology of Fresh Waters, 3rd ed. (Blackwell Science, Oxford, UK, 1998).

A.S. Mungur, R.B.E. Shutes, D.M. Revitt and M.A. House, "An assessment of metal reduction by laboratory-scale wetlands", Wat. Sci. Tech., 35 (5), 125–133 (1997).

R.W. Nairn and W.J. Mitsch, "Phosphorus removal in created wetland ponds receiving river overflow", Ecol. Eng., 14 (1–2), 107–126 (2000).

A.C. Norrstrom and G. Jacks, "Concentration and fractionation of heavy metals in roadside soils receiving de-icing salts", Sci. Tot. Env., 218 (2–3), 161–174 (1998).

P.M. Nuttal, A.G. Boom and M.R. Rowell, Review of the Design and Management of Constructed Wetlands. Construction Industry Research and Information Association (CIRIA), CIRIA Report 180 (CIRIA, London, UK, 1997).

H. Obarska-Pempkowiak and K. Klimkowska, "Distribution of nutrients and heavy metals in a constructed wetland system", Chemosphere, 39 (2), 303–312 (1999).

G. Onkal-Engin, I. Demir and S.N. Engin, "Determination of the relationship between sewage odor and BOD by neural networks", Environ. Model. Softw., 20 (7), 843–850 (2005).

W. Ostendorp, "'Die-back' of reeds in Europe – a critical review of literature", Aquat. Bot., 35 (1), 5–26 (1989).

J. Overbeck, Qualitative and quantitative assessment of the problem, In: S.E. Jørgensen and R.A. Vollenweider (eds.), Guidelines for Lake Management: Principles of Lake Management, Vol. 1 (International Lake Environment Committee, United Nations Environment Programme, 1988) pp. 19–36.

P.-F. Pai and W.-C. Hong, "Support vector machines with simulated annealing algorithms in electricity load forecasting", Energ. Convers. Manage., 46 (17), 2669–2688 (2005).

S.J. Park and T.I. Yoon, "Weighted coagulation with glass and diatomite for stormwater treatment and sludge disposal", Environ. Eng. Sci., 20 (4), 307–317 (2003).

T.W. Parr, Factors affecting reed (*Phragmites australis*) growth in UK reed bed treatment systems, In: P.F. Cooper and B.C. Findlater (eds.), Constructed Wetlands in Water Pollution Control (Pergamon Press, Oxford, UK, 1990) pp. 67–76.

Perkin Elmer, Perkin Elmer 1100 Series Atomic Absorption Spectrophotometer (Perkin Elmer, London, UK, 1982).

J. Perkins and C. Hunter, "Removal of enteric bacteria in a surface flow constructed wetland in Yorkshire, England", Wat. Res., 34 (6), 1941–1947 (2000).

T.J.R. Pettersson, "FEM-modelling of open stormwater detention ponds", Nord. Hydrol., 28 (4–5), 339–350 (1997).

C.R. Picard, L.H. Fraser and D. Steer, "The interacting effects of temperature and plant community type on nutrient removal in wetland microcosms", Biores. Technol., 96 (9), 1039–1047 (2004).

M.L. Pinney, P.K. Westerhoff and L. Baker, "Transformations in dissolved organic carbon through constructed wetlands", Wat. Res., 34 (6), 1897–1911 (2000).

C.D. Preston and J.M. Croft, Aquatic Plants in Britain and Ireland (Harley Books, Colchester, England, 1997).

K.R. Reddy, R.H. Kadlec, E. Flaig and P.M. Gale, "Phosphorus retention in streams and wetlands: a review", Crit. Rev. Env. Sci. Tech., 29 (1), 83–146 (1999).

E.W. Reiche, "Modelling water and nitrogen dynamics on catchment scale", Ecol. Model., 75–76, 371–384 (1994).

M.J. Richardson and F.G. Whoriskey, "Factors influencing the production of turbidity by goldfish (*Carassius auratus*)", Can. J. of Zool., 70 (8), 1585–1589 (1992).

S.M. Ross, Toxic Metals in Soil-Plant Systems (John Wiley and Sons, Sussex, UK, 1994).

J.W.M. Rudd, C.A. Kelly, V. St. Louis, R.H. Hesslein, A. Furutani and M.H. Holoka, "The role of sulfate reduction in long term accumulation of organic and inorganic sulfur in lake sediments", Limnol. Oceanogr., 31 (6), 1281–1291 (1986).

J. Ruiz-Jimenez, F. Priego-Capote, J. Garcia-Olmo and M.D. Luque de Castro, "Use of chemometrics and mid infrared spectroscopy for the selection of extraction alternatives to reference analytical methods for total fat isolation", Anal. Chim. Acta, 525 (2), 159–169 (2004).

J.J. Sansalone, "Adsorptive infiltration of metals in urban drainage – media characteristics", Sci. Tot. Env., 235 (1–3), 179–188 (1999).

K. Sasaki, T. Ogino, Y. Endo, and K. Kurosawa, "Field studies on heavy metal accumulation in a natural wetland receiving acid mine drainage", Mater. Trans., 44 (9), 1877–1884 (2003).

W.H. Schlesinger, Biogeochemistry: An Analysis of Global Change (Academic Press, San Diego, USA, 1991).

T.G. Schmitt, M. Thomas and N. Ettrich, "Analysis and modelling of flooding in urban drainage systems", J. Hydro., 299 (3–4), 300–311(2004).

M. Scholz, "Case study: design, operation, maintenance and water quality management of sustainable stormwater ponds for roof runoff", Biores. Technol., 95 (3), 269–279 (2003).

M. Scholz, "Treatment of gully pot effluent containing nickel and copper with constructed wetlands in a cold climate", J. Chem. Tech. Biotech., 79 (2), 153–162 (2004a).

M. Scholz, "Stormwater quality associated with a silt trap (empty and full) discharging into an urban watercourse in Scotland", Intern. J. Environ. Studies, 61 (4), 471–483 (2004b).

M. Scholz and R.J. Martin, "Control of bio-regenerated granular activated carbon by spreadsheet modelling", J. Chem. Tech. Biotechnol., 71 (3), 253–261 (1998a).

M. Scholz and R.J. Martin, "Biological control in granular activated carbon beds", Internat. Rev. Hydrobiol., 83 (special issue), 657–664 (1998b).

M. Scholz and M. Trepel, "Water quality characteristics of vegetated groundwater-fed ditches in a riparian peatland", Sci. Tot. Env., 332 (1–3), 109–122 (2004).

M. Scholz and J. Xu, "Performance comparison of experimental constructed wetlands with different filter media and macrophytes treating industrial wastewater contaminated with lead and copper", Biores. Technol., 83 (2), 71–79 (2002).

M. Scholz and S. Zettel, "Stormwater quality associated with a full silt trap discharging into urban watercourses", J. Chart. Inst. Wat. Environ. Manage., 18 (4), 226–229 (2004).

M. Scholz, P. Höhn and R. Minall, "Mature experimental constructed wetlands treating urban water receiving high metal loads", Biotechnol. Prog., 18 (6), 1257–1264 (2002).

M. Scholz, R. Morgan and A. Picher, "Stormwater resources development and management in Glasgow: two case studies", Intern. J. Environ. Studies, 62 (3), 263–282 (2005).

E.A. Seaman, "Observations on *Carassius auratus* Linneaus harvesting *Potamogeton foliosus* Raf. in a small pond in north Virginia", Fisheries, 4 (3), 24–25 (1979).

SEPA, Protecting the Quality of Our Environment – Sustainable Urban Drainage: An Introduction, Scottish Environmental Protection Agency (SEPA) (Stationary Office, London, UK, 1999).

SEPA, SUDS Advice Note – Brownfield Sites (Scottish Environmental Protection Agency (SEPA), Stirling, UK, 2003).

R.E. Shutes, D.M. Revitt, L.N.L. Scholes, M. Forshaw and B. Winter, "An experimental constructed wetland system for the treatment of highway runoff in the United Kingdom", Wat. Sci. Tech., 44 (11–12), 571–578 (2001).

F. Simpson and M. Chapman, "Comparison of urban governance and planning policy – east looking", West. Cities, 16 (5), 353–364 (1999).

R.L. Smith, Ecology and Field Biology, 3rd ed. (Harper and Row, New York, USA, 1980).

C.S. Smith and M.S. Adams, "Phosphorus transfer from sediments to Myriophyllum spicatum", Limnol. Oceanogr., 31 (6), 1312–1321 (1986).

D.J. Spieles and W.J. Mitsch, "The effects of season and hydrologic and chemical loading on nitrate retention in constructed wetlands: a comparison of low- and high-nutrient riverine systems", Ecol. Eng., 14 (1–2), 77–91 (2000).

H.G. Spratt and M.D. Morgan, "Sulfur cycling in a cedar dominated freshwater wetland", Limnol. Oceanogr., 35 (7), 1586–1593 (1990).

W. Stumm and J.J. Morgan, Aquatic Chemistry: Chemical Equilibria and Rates in Natural Waters, 3rd ed. (John Wiley and Sons, New York, 1996).

G. Sun, K.R. Gray, A.J. Biddlestone and D.J. Cooper, "Treatment of agricultural wastewater in a combined tidal flow downflow reed bed system", Wat. Sci. Tech., 40 (3), 139–146 (1999).

K.H. Tan, Soil Sampling, Preparation and Analysis (Marcel Dekker, New York, USA, 1996).

C.C. Tanner, "Plants for constructed wetland treatment systems – a comparison of the growth and nutrient uptake of eight emergent species", Ecol. Eng., 7 (1) 59–83 (1996).

G. Tchobanoglous, F.L. Burton and H.D. Stensel, Wastewater Engineering: Treatment and Reuse, 4th ed. (Metcalf and Eddy, McGraw Hill Companies, New York, USA, 2003).

M. Trepel, "Spatiotemporal simulation of water- and nitrogen dynamics as a tool in fen restoration", Int. Peat J., 9, 45–52 (1999).

M. Trepel, "Gedanken zur zukünftigen Nutzung schleswig-holsteinischer Niedermoore", Die Heimat, 108 (11–12), 186–194 (2001) (in German).

M. Trepel and W. Kluge, "Ecohydrological characterisation of a degenerated valley peatland in Northern Germany for use in restoration", J. Nat. Conserv., 10, 155–169 (2002).

M. Trepel and W. Kluge, "WETTRANS: a flow-path-oriented decision support system for the assessment of water and nitrogen exchange in riparian wetlands", Hydrol. Proc., 18, 357–371 (2004).

M. Trepel and L. Palmeri, "Quantifying nitrogen retention in surface flow wetlands for environmental planning at the landscape-scale", Ecol. Eng., 19 (2), 127–140 (2002).

M. Trepel, B. Holsten, J. Kieckbusch, I. Otten and F. Pieper, Influence of macrophytes on water level and flood dynamics in a riverine wetland in Northern Germany, In: Proceedings of the International Conference EcoFlood - Towards Natural Flood Reduction Strategies, Warshaw, 6–13 September 2003 (Institute for Land Reclamation and Grassland Farming, Raszyn, Poland, 2003).

M.K. Turner, "Engineered reed-bed systems for wastewater treatment", Trends in Biotech., 13 (7), 248–252 (1995).

I. Turok and N. Bailey, "Twin track cities? Competitiveness and cohesion in Glasgow and Edinburgh", Progr. Plang., 62 (3), 135–204 (2004).

T.W. Tyson, Best Management Practices for Animal Feeding Operations, ANR-1188 (Alabama Cooperative Extension Systems, USA, 2000).

Umweltbundesamt, Daten zur Umwelt: Der Zustand der Umwelt in Deutschland 2000 (Erich Schmidt Verlag, Berlin, Germany, 2001) (in German).

United States Army Corps of Engineers, Wetlands Engineering Handbook - ERDC/EL TR-WRP-RE-21 (United States Army Engineer Research and Development Center Cataloging-in-Publication Data, 2000).

M. van der Aa, M. Trepel, P.F.M. van Gaans, W. Bleuten and W. Kluge, "Modelling water flow and fluxes of a valley mire for use in restoration", Landnutzung und Landentwicklung, 42, 72–78 (2001) (in German).

A.G. van der Valk and C.B. Davis, Primary production of prairie glacial marshes, In: R.E. Good, D.F. Whigham and R.L. Simpson (eds.) Freshwater Wetlands: Ecological Processes and Management Potential (Academic Press, New York, USA, 1978) pp. 21–37.

V. Vapnik, The Nature of Statistics Learning Theory (Springer Verlag, New York, USA, 1995).

F. Verdenius and J. Broeze, "Generalized and instance-specific modeling for biological systems", Environ, Model. Softw., 14 (5), 339–348 (1999).

J. Vesanto, J. Himberg, E. Alhoniemi and J. Parhankangas, "Self-organizing Map in Matlab: the SOM Toolbox", *Proceedings of the Matlab DSP Conference*, Espoo, Finland, Nov. 1999, pp. 35–40. Software available at http://www.cis.hut.fi/projects/somtoolbox/.

E.L. Villarreal, A. Semadeni-Davies and L. Bengtsson, "Inner city stormwater control using a combination of best management practices", Ecolog. Eng., 22 (4–5), 279–298 (2004).

J. Vymazal, "The use of sub-surface constructed wetlands for wastewater treatment in the Czech Republic: 10 years experience", Ecol. Eng., 18 (5), 633–646 (2002).

J. Vymazal and P. Krasa, "Distribution of Mn, Al, Cu and Zn in a constructed wetland receiving municipal sewage", Wat. Sci. Tech., 48 (5), 299–305 (2003).

J. Vymazal, H. Brix, P.F. Cooper, M.B. Green and R. Haberle, Constructed Wetlands for Wastewater Treatment in Europe (Backhuys, Leiden, The Netherlands, 1998).

N. Wang and W.J. Mitsch, "A detailed ecosystem model of phosphorus dynamics in created riparian wetlands", Ecol. Model., 126 (2–3), 101–130 (2000).

H. Werner and M. Obach, "New neural network types estimating the accuracy of response for ecological modelling", Ecol. Model., 146 (1–3), 289–298 (2001).

A. Wießner, U. Kappelmeyer, P. Kuschk and M. Kästner, "Influence of the redox condition dynamics on the removal efficiency of a laboratory-scale constructed wetland", Wat. Res., 39 (1), 248–256 (2004).

A. Wild, Soils and the Environment (Cambridge University Press, Cambridge, UK, 1993).

S. Wilson, R. Bray and P. Cooper, Sustainable Drainage Systems. Hydraulic, Structural and Water Quality Advice. Construction Industry Research and Information Association (CIRIA) Report C609 (CIRIA, London, UK, 2004).

R. Wisskirchen and H. Häupler, Standartliste der Farn- und Blütenpflanzen Deutschlands, Bundesamt für Naturschutz (Verlag Eugen Ulmer, Stuttgart, 1998) (in German).

T.S. Wood and M.L. Shelley, "A dynamic model of bioavailability of metals in constructed wetland sediments", Ecol. Eng., 12 (3–4), 231–252 (1999).

D.A. Wrubleski, H.R. Murkin, A.G. van der Valk and J.W. Nelson, "Decomposition of emergent macrophyte roots and rhizomes in a northern prairie marsh", Aquat. Bot., 58 (2), 121–134 (1997).

X.Y Wu and W.J. Mitsch, "Spatial and temporal patterns of algae in newly constructed freshwater wetlands", Wetlands, 18 (1), 9–20 (1998).

Y.Q. Zhao, G. Sun and S.J. Alen, "Purification capacity of a highly loaded laboratory scale tidal flow reed bed system with effluent recirculation", Sci. Tot. Env., 330 (1–3), 1–18 (2004).

P.Q. Zheng and B.W. Baetz, "GIS-based analysis of development options from a hydrology perspective", J. Urban Plng. and Devel., 125 (4), 164–180 (1999).

Index